Introduction to Electronic Warfare Modeling and Simulation

Introduction to Electronic Warfare Modeling and Simulation

David L. Adamy

SCITECH
PUBLISHING, INC.

SciTech Publishing, Inc
Raleigh, NC
www.scitechpub.com

SCITECH
PUBLISHING, INC.

First Artech House, Inc. edition 2003
First SciTech Publishing, Inc. edition 2006

Printed in the United States of America

ISBN: 1-891121-62-6
ISBN 13: 978-1-891121-62-3

Printed in the United States of America.

SciTech Publishing
911 Paverstone Drive, Suite B
Raleigh, NC 27615
Phone: 919-847-2434
Fax: 919-847-2568
www.scitechpub.com

Library of Congress Cataloging-in-Publication Data
Adamy, David
    Introduction to electronic warfare modeling and simulation/David L.
    Adamy.
        p. cm.—(SciTech Publishing, Inc.)
    Includes bibliographical references and index.
    ISBN 1-891121-62-6
    1. Electronics in military engineering—Mathematical models.
    2. Electronics in military engineering—Computer simulation.
    I. Title.   II. Series.
    UG 485 .A3323   2002
    355.4'01'13—dc21   2002027779

Contents

Preface

Modeling and simulation, when done well, is an insightful pursuit. It starts with an understanding of what is really happening, then proceeds to the mathematical modeling of what is happening, and then—if necessary—to the recreation of that reality from some point of view.

In electronic warfare (EW) modeling and simulation, the reality is an array of threats—emitting threat-related signals—received by EW systems. Models capture the action and either determine what will happen if various trade-offs are made or reproduce that threat environment as it is seen from one of many possible points of view.

To do EW modeling and simulation well, you need to be able to view the tactical situation from unusual points of view. For example, before you can simulate a set of signals to make a receiver think it is in a big war, you need to be able to figure out what the big war looks like when you are a receiver looking out through a front-panel connector.

This book is designed to help you develop the ability to look at a tactical situation from the point of view of an antenna, a receiver, or an operator (perhaps in an aircraft being shot at while flying upside down).

Although there is much information available about specific models, simulators, and simulations, providing such information is not within the scope or intent of this book. The reader is referred to the specific product manufacturers. Periodic surveys in the *Journal of Electronic Defense* (www. jedonline.com) list the latest EW simulators, along with their capabilities and the organizations that produce them. An excellent way to begin is to call a company and ask for application engineering, marketing, or customer support. Upon request, companies will typically send out literature that includes a wealth of unclassified information about their products. They will also be able to steer you to the current government policymakers in the modeling and simulation fields.

EW is a large field, dealing with the electromagnetic spectrum from just above dc to just above daylight. This means that the field of EW modeling and simulation is just as broad. In order to maximize the usefulness of this book, its focus is on the radio frequency spectrum, with primary emphasis on HF to microwave. Further, it concentrates on EW applications and techniques rather than detailed propagation models (more complex than line-of-sight propagation).

The book is designed for use by anyone requiring knowledge of EW modeling and simulation; it presumes that readers have a general technical background and knowledge of basic algebra. The text begins with an overview of EW and explains all necessary math above algebra.

Welcome to the world of EW modeling and simulation. I hope you will enjoy it as much as I have.

Acknowledgments

This book is dedicated to the professionals who have developed the art and science of electronic warfare modeling and simulation. Although theirs is not a glory field, they serve nobly behind the scenes. Their expertise and dedication allow for the development of equipment that can do what it must when it goes into harm's way—and the effective training of the people who take it there.

Particular thanks goes to the following colleagues who took the time to review the draft of this book. They represent a combined two centuries of relevant professional experience. Each contributed expertise and insight about the field and about the needs of the people who follow it.

Mr. Dave Barton
Mr. Bob Dalton
Mr. Joe DiGiovanni
Prof. Fred Levien
Mr. Paul McGillick
Col. Linda Palmer
Mr. Bill Shellenberger
Dr. Ed Wischmeyer

1

Introduction

Electronic warfare (EW) simulation is a serious game of make-believe. A situation is artificially created so that equipment can be tested and operators can be trained under realistic conditions—without both the expense and danger associated with training in the real world. EW engagements are typically complex, with many threat emitters seen in constantly changing relationships as either the threat platforms or the EW-protected platform maneuver. The operation of equipment and the performance required of operators cannot be adequately tested under static conditions with only one threat emitter present. This has led to the development of simulators that can simulate many threats grouped and maneuvered in realistic ways.

1.1 Simulation

Simulation is the creation of an artificial situation or stimulus that causes an outcome to occur as though a corresponding real situation or stimulus were present.

Any of several levels of simulation can be applied to any kind of stimulus. For example, in a flight-simulator computer program, the computer screen shows what a pilot would see through the windshield and on the instrument panel of a particular type of aircraft. The view and the instrument readouts change in response to the manipulation of simulated controls. In a more elaborate flight simulator, the pilot sits in a fully simulated cockpit, which tilts and moves to simulate the g forces created by aircraft movement. The visual cues may include a full hemispherical view from the cockpit (including attacking enemy aircraft), and the cockpit sounds will also be provided. All of these displays are interactive, changing in response to control manipulation by the pilot and to selected external events (for example, engine failure).

There are also rifle-firing simulators in which a subject holds what seems to be an actual rifle. On a screen is a firing range with targets the subject is to engage. When the rifle is fired, the subject hears the sound of the rifle firing and feels the rifle's recoil. The direction that the rifle is pointed is coordinated with the computer-generated screen and the strikes of the shots fired by the student are shown.

These are only two examples of the many kinds of simulators that are used for training in many fields. Simulators can also be used to develop and evaluate tactics to deal with almost any kind of external event affecting almost any kind of equipment in almost any field.

1.1.1 Simulation in the EW Field

Due to the nature of EW systems, EW simulation often involves the creation of signals like those generated by an enemy's electronic assets. These artificial signals are used to train operators or to evaluate the performance of EW systems and subsystems. The EW simulation field also involves the prediction of the performance of enemy electronic assets or the weapons they control.

Through simulation, an operator or piece of EW equipment can be caused to react as though one or more threat signals were present and doing what they would be doing during a military encounter. Typically, the simulation involves interactive changing of the threat situation or the way it is processed as a function of what the operator or equipment does in response to the perceived presence of the threat signals.

EW simulation is often added to military vehicle simulators (for example, flight simulators) so that the displays of EW equipment respond appropriately to the flight situation that is presented to the subject through visual, motion, and sound cues.

On the other hand, a simulator may simply supply electronic inputs to a receiver system, subsystem, or individual chassis. The inputs provide the equipment with what it would "see" in the situation in which it is designed to operate.

1.1.2 Modeling

Modeling, in the sense used here, is the mathematical characterization of an item or a process. It is an integral part of simulation, because anything that is simulated must first be defined mathematically. In EW simulation, we model the performance characteristics of hardware, systems, and platforms. We also model engagements between platforms (for example, an enemy aircraft and a friendly aircraft or ship) and between systems (for example, an enemy radar and a friendly jammer). The first step in developing any model is a conceptual analysis of what is happening; after that a mathematical model can be constructed.

Modeling is typically performed in computers, but can also be performed with pencil and paper. The big advantage of computer modeling is that the details of the model are captured and the process can be run multiple times with controlled granularity and dependable repeatability.

1.2 EW Simulation Approaches

Simulation is often divided into three subcategories: computer simulation, operator interface simulation, and emulation. Computer simulation is also called *modeling*. Operator interface simulation is often simply called *simulation*. This can cause confusion, because the same term is commonly used to define both the simulation field and this particular approach. Emulation involves the creation of actual signals. All three approaches are employed for either training or test and evaluation (T&E). Table 1.1 shows the frequency of the use of each subcategory for each purpose.

1.2.1 Modeling

Computer simulation (or modeling) is done in a computer using mathematical representations of friendly and enemy assets and sometimes operator responses—evaluating how they interact with each other. In modeling, neither signals nor representations of tactical operator controls and displays are generated. The purpose is simply to evaluate the interaction of equipment

Table 1.1
Simulation Approaches and Purposes

Simulation Purpose	Simulation Approach		
	Modeling	*Simulation*	*Emulation*
Training	Commonly	Commonly	Sometimes
T&E	Sometimes	Seldom	Commonly

and tactics that can be mathematically defined. Modeling is extremely useful for the evaluation of strategies and tactics. A situation is defined, and each of several approaches is implemented. The outcomes are then compared. It is important to note that any simulation or emulation must be based on a model of the interaction between an EW system and a threat environment, as shown in Figure 1.1.

1.2.2 Simulation

Operator-interface simulation refers to the generation of operator displays and the reading of operator controls in response to a situation that has been modeled and is proceeding, but without the generation of any actual signals. The operator sees computer-generated displays and hears computer-generated audio just as though the EW system he or she is operating were in a specified tactical situation. The computer reads the operator's control responses and modifies the displayed information accordingly. If the operator's control actions would be expected to modify the tactical situation, this is also reflected in the display presentations.

In some applications, operator-interface simulation is achieved by driving the actual system displays from a simulation computer. Switches are read as binary inputs, and analog controls (for example, a rotated volume control) are usually attached to shaft encoders to provide a computer-readable knob position.

The other approach is to generate artificial representations of the system displays on one or more computer screens. Displays are represented as

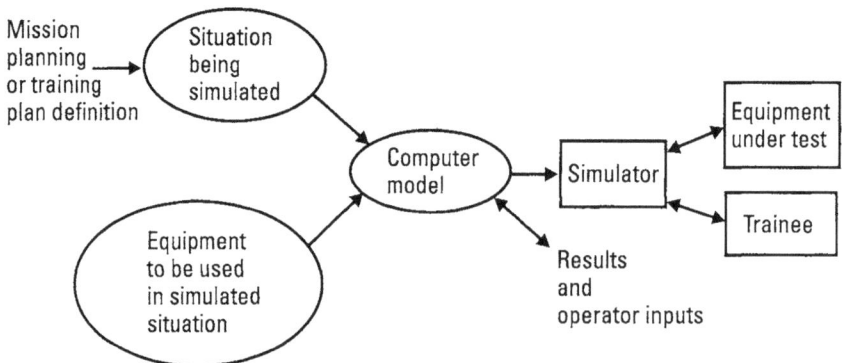

Figure 1.1 EW system and threat environment model interaction.

pictures of the system displays, usually including a portion of the instrument panel in addition to the actual dials or CRT screens. Controls are pictorially shown on the computer screen and operated by use of a mouse or touchscreen feature.

1.2.3 Emulation

Where any part of the actual system is present (except, perhaps, computerdriven controls and displays), the emulation approach is used. Emulation involves the generation of signals in the form that they would have at the point where they are injected into the system. Although the emulation approach can be used for training, it must almost always be used for the T&E of systems or subsystems.

As shown in Figure 1.2, emulated signals can be injected at many points in the system. The trick is to make the injected signal look and act as if it has come through the whole system—in the simulated tactical situation. Another important point is that anything that happens downstream of the injection point (for example, automatic gain control or operator control actions) may have an impact on the signal arriving at the injection point. If so, the injected signal must be appropriately modified.

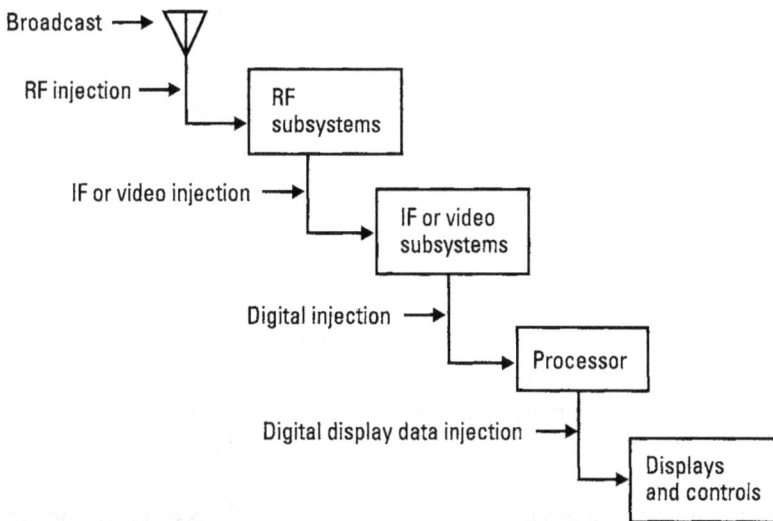

Figure 1.2 Points of injection of emulated signals in EW system.

1.3   Simulation for Training

Simulation for training exposes students to experiences (in a safe and controlled way) that allow them to learn or practice various types of skills. In EW training, this most often involves the experiencing of enemy signals in the way they would be encountered if the trainee were at an operating position in a military situation. EW simulation is often combined with other types of simulation to provide a full training experience. For example, a cockpit simulator for a particular aircraft may include EW displays that react as though the aircraft were flying through a hostile electronic environment. Training simulation usually allows an instructor to observe what the student sees and how he or she responds. Sometimes, the instructor can play back the situation and responses as part of the debriefing after a training exercise—a powerful learning experience.

1.4   Simulation for T&E

Simulation for equipment T&E involves making a piece of equipment think it is doing the job for which it was designed. This can be as simple as generating a signal with the characteristics a sensor is designed to detect. It can be as complex as generating a realistic signal environment containing all of the signals a full system will experience as it moves through a lengthy engagement scenario. Further, that environment may vary in response to a preprogrammed or operator-selected sequence of control and movement actions by the system being tested. It is distinguished from training simulation in that its purpose is to determine how well the equipment works rather than to impart skills to operators.

1.5   Electronic Point of View

An important concept in simulation is point of view. While the planner and designer of a simulation know the whole situation (because they have postulated it), the simulation must represent what the equipment or trainee for whose benefit the simulation is created would experience in the postulated situation. This depends on the sensors available and their operating modes (see Figure 1.3).

For example, a receiver "sees" the world through its input connector. It does not care where the emitter is located or what mode it is in—the receiver can only see what its antenna outputs to it. If it is connected to a scanning

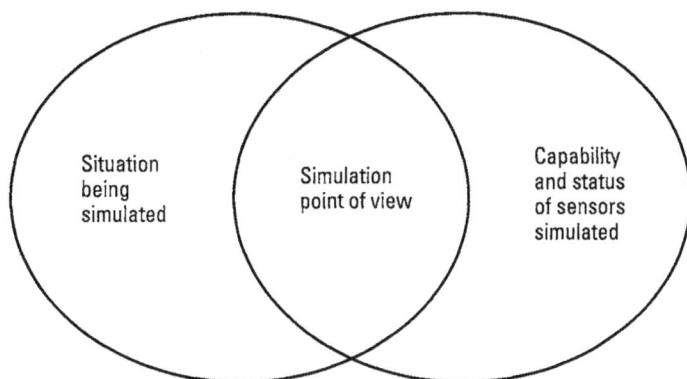

Figure 1.3 Simulation point of view.

narrow-beam antenna that is pointed away from an emitter, the receiver will not know that the emitter has just come on the air. Likewise, if an operator has not tuned his or her receiver to the frequency of the signal from the emitter, the operator will not experience that emitter.

1.6 Fidelity in EW Simulation

Fidelity is an important consideration in the design or selection of an EW simulator. The fidelity of the model and of the data presented to systems and operators must be adequate to the task. In training simulation, the fidelity must be adequate to prevent the operator from noticing any inaccuracies caused by the simulation (or at least to avoid distortions that will interfere with the training objectives). In T&E simulation, the fidelity must be adequate to provide injected signal accuracy better than the perception threshold of the tested equipment. Both of these issues will be discussed in detail in later chapters. As shown in Figure 1.4, the cost of a simulation rises—often exponentially—as a function of the fidelity provided. The curve in the figure shows a sharp cutoff of value versus cost to illustrate the point that the value of the fidelity does not increase once the perception level of the trainee or tested equipment is reached. Any additional fidelity is a waste of money or of equipment complexity (which typically reduces its reliability or increases its requirement for maintenance). However, it must be clarified that the cutoff is not always this sharp; judgment and the consideration of many factors in the design of the tested equipment and the employment of the simulator are required.

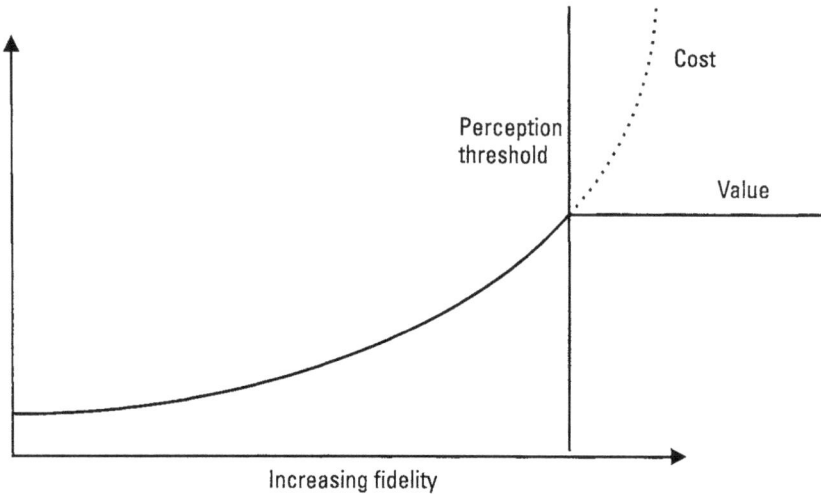

Figure 1.4 Cost of simulation and fidelity.

A related point is that cost also increases exponentially with the amount of human action to be accounted for. For example, it is generally much simpler to reproduce the predictable mechanical actions of a radar than the less predictable decisions made by human operators.

The specifics of the required fidelity will be detailed in later chapters discussing modeling, simulation for training, and simulation for T&E.

1.7   The Tactical Big Picture

EW modeling and simulation starts with a military situation with EW implications that must be modeled or simulated. As shown in Figure 1.5, the big picture of what is happening typically involves enemy weapons. There are electronic signals associated with the deployment and employment of these weapons. The signals include those generated by communications and radar transmitters. The combination of all of these signals comprises the threat signal environment.

As weapons move, the locations of transmitters on weapon platforms also move. Signals from related transmitters also change in response to (or in preparation for) weapon movements. Finally, the location and movement of friendly assets (targeted by enemy weapons) cause enemy weapons to move or to change operating modes in order to engage their targets. There are also

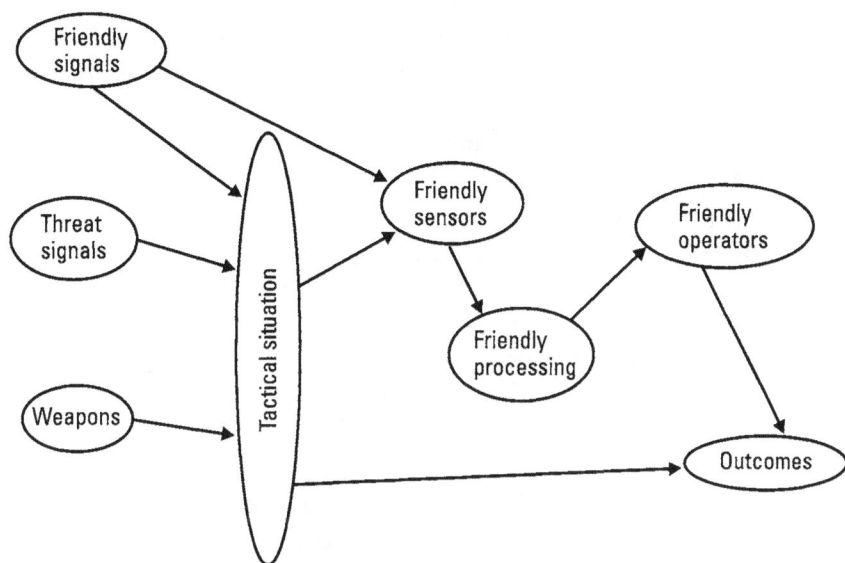

Figure 1.5 Tactical big picture.

signal transmissions from these friendly assets. Since the model of this interaction has an observational point of view, (for example, a friendly receiver or an operator), the location of the observer may need to be changed to reflect the motion of modeled friendly or enemy assets.

Taken together, the combined enemy and friendly transmissions reflect the tactical situation. Receivers associated with friendly EW systems collect and analyze enemy signals to derive necessary information about enemy weapons and other assets, and also deal with all of the other signals in the signal environment.

In Figure 1.5, friendly sensors are shown observing the tactical situation to acknowledge that there are various types of friendly sensors, including radars, infrared, and electro-optical sensors, as well as radio frequency receivers. The outputs of these sensors are processed and usually evaluated or interpreted by operators who cause actions to be taken. There are also automatic responses to perceived threats in the tactical situation.

As a result of the tactical situation and the automatic or manually invoked EW responses, outcomes occur. These outcomes may be the destruction of friendly assets by the enemy, the defeat of enemy weapons, or the destruction of enemy weapons.

The big picture relates to EW modeling and simulation as follows:

- An EW model might include an enemy and friendly asset laydown along with a set of required responses of both enemy and friendly assets to actions by the other side. Then, the model would be run to determine the outcomes under varying conditions and with different EW tactics and EW assets.

- An EW simulation might build on the model to generate displays that would be seen by friendly operators while sensing operator control actions and reflecting those actions in the tactical situation. This will in turn be reflected in the operator displays. The simulation might also go on to determine the outcomes—including the effects of operator actions. The purpose of this simulation is, of course, to train and evaluate the skills of operators.

- An EW simulation might also emulate the input signals to the friendly sensors as a function of the tactical environment. Then the sensor or processor outputs can be evaluated, either manually or automatically, to determine how well the sensors (or processors) can be expected to perform under various enemy action scenarios.

- A threat scenario is a series of interactions with threat signals as seen by some friendly sensor.

- A simulation can include man-in-the-loop features. That is, humans can be involved in the simulation as friendly or expert hostile operators to make the operator training or equipment testing more realistic.

1.8 Simulation Versus Life Cycle

Modeling and simulation play a role in many aspects of the development, manufacture, and tactical use of EW systems over their entire life cycle. At the beginning, computer modeling helps establish requirements and concept development. It also assists in tradeoff analysis studies. During the development stages, signal injection into modules, units, subsystems, and the whole system predict later operational performance. Realistic early testing can have a powerful effect on the total development cost. During operational evaluation, high-power radiating simulators provide realistic evaluation of system performance in the actual environments to be encountered over the operational lifetime. Operator-interface simulation is invaluable in the training of operators over the operational lifetime of EW systems.

2

Overview of EW

EW is the art and science of denying enemy forces the use of the electromagnetic spectrum while preserving its use for friendly forces. EW is a significant force multiplier because it reduces friendly losses by defeating or reducing the effectiveness of enemy weapons.

As shown in Figure 2.1, EW is commonly divided into three subfields. Electronic support (ES) involves receiving enemy signals to identify and locate threat emitters and to help determine the enemy's force structure and deployment. Electronic attack (EA) involves measures taken to defeat enemy electronic assets. It includes jamming, chaff and flares, directed energy weapons, and antiradiation missiles. Electronic protection (EP) comprises countermeasures to enemy electronic attack. All of these subfields apply at radar, communication, infrared, and laser operating frequencies.

This chapter deals with the signals, equipment, deployment, and tactics associated with all aspects of EW. As indicated by the chapter title, this discussion provides an overview of EW. It is limited to providing enough information to support our later discussion of EW modeling and simulation, and focuses primarily on the aspects important to modeling and simulation. The following books are recommended to provide the reader with more depth and breadth in the field:

- David Adamy. *EW101: A First Course in Electronic Warfare*. Norwood, MA: Artech House, 2001 (www.artechhouse.com);

```
            ┌─────────────┐
            │ Electronic  │
            │  warfare    │
            └─────────────┘
                   │
      ┌────────────┼────────────┐
┌───────────┐ ┌───────────┐ ┌───────────┐
│ Electronic│ │ Electronic│ │ Electronic│
│  support  │ │  attack   │ │ protection│
└───────────┘ └───────────┘ └───────────┘
```

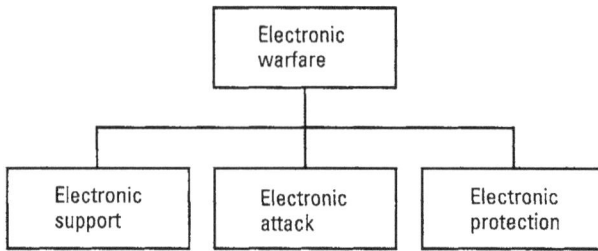

Figure 2.1 Structure of EW.

- David B. Hoisington. *Electronic Warfare.* Monterey, CA: Naval Postgraduate School, 1980 (available from Lynx Publishing at www.lynxpub.com);

- D. Curtis Schleher. *Introduction to Electronic Warfare.* Norwood, MA: Artech House, 1986;

- D. Curtis Schleher. *Electronic Warfare in the Information Age.* Norwood, MA: Artech House, 1999;

- Filippo Neri. *Introduction to Electronic Defense Systems, Second Edition,* Norwood, MA: Artech House, 2001.

We turn now to a discussion of the signals associated with EW (radar and communication) and the three major subfields.

2.1 Radar

Radar was developed to detect and track aircraft and ships so that they could be attacked. Now there are a wide variety of radars for both peaceful and warlike purposes. For example, radars track aircraft for air-traffic control, determine weather conditions, map ground contours, track ships or surface vehicles, and detect people moving over terrain.

2.1.1 Basic Radar Function

A radar has both a transmitter and a receiver. The transmitted signal is reflected by anything it strikes and those reflected signals are received and processed. The transmitter and receiver are usually (but not necessarily) colocated, and may share a common antenna. Figure 2.2 shows the function of a simple radar.

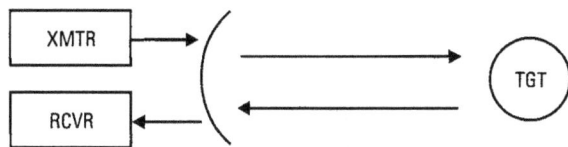

Figure 2.2 Simple radar diagram.

The transmitter (XMTR) transmits a signal to a common antenna (a parabolic dish is very typical). The antenna is almost always highly directional so that the transmitted signal is constrained to a narrow angular segment (called the *beam width*) so that it illuminates only a single target (TGT). In Section 5.1, the various types of antennas and their characteristics will be discussed in detail. Since the receiver (RCVR) uses the same antenna, it receives reflections in its main beam from the illuminated target. The azimuth and elevation angle of the antenna is known, so the angular location of the target can be readily determined.

The distance to the target is determined by measuring the time between the transmission of the signal and the reception of the signal reflected from the target. Since the signal always travels at the speed of light, the round trip propagation time allows a very accurate measurement of the distance to the target. The equation for the distance to the target is

$$d = ct / 2$$

where

d = the distance to the target (m);

c = the speed of light (3×10^8 m/s);

t = round-trip time (sec).

Another way to look at this is that radio waves travel 1 foot (or 1/3 meter) per nanosecond. Since a radar deals with the round-trip time, the distance is half a foot for each nanosecond delay of the returned signal.

If the location of the target is unknown, the radar beam is rotated in some pattern to cover the angular space expected to contain the target. This movement of the antenna is called the *antenna scan.* The antenna can either scan continuously or lock on to the target, depending on the radar's mission. EW receivers often analyze this scanning pattern as one of the ways to identify the type of the radar or its operating mode. The appearance of many types of scans to an EW receiver will be discussed in Chapter 5.

Many radars also measure the rate of change of the range to the target. By measuring the difference between the frequency of the transmitted and received signals (called the *Doppler shift*), a radar determines the rate of change of distance (or the component of relative target velocity in the direction of the radar). The Doppler shift formula (for this type of radar) is

$$\Delta d / \Delta t = 0.5 \, c \left(\Delta F / F \right)$$

where

$\Delta d/\Delta t$ = the rate of change of distance to the target;

c = the speed of light;

ΔF = the difference frequency;

F = transmitted frequency.

The one-half factor is because the signal has a two-way path, so there is a Doppler shift for the transmitted signal path and another for the reflected signal path. It should be noted that a receiver located on the target would only be affected by a single Doppler shift.

Radars can be either monostatic or bistatic. A monostatic radar has the transmitter and receiver at the same location. In bistatic radars, the transmitter and receiver are at different locations. In this discussion, monostatic radar configuration is assumed unless otherwise indicated. There are, however, several important applications of bistatic radar that will be discussed as we proceed.

2.1.2 Radar Modulations

There are three basic types of radar modulation: pulse, continuous wave, and pulse Doppler.

Pulse Radar

The most common type of radar modulation is a stream of pulses. Pulses are short transmissions, as shown in Figure 2.3. Their duration (called the *pulse width*) is typically from part of a microsecond to several microseconds. The interval between pulses [called the *pulse repetition interval* (PRI)] is typically a fraction of a millisecond to several milliseconds. The inverse of the PRI is the pulse repetition frequency (PRF). Pulse modulation has several advantages for radars:

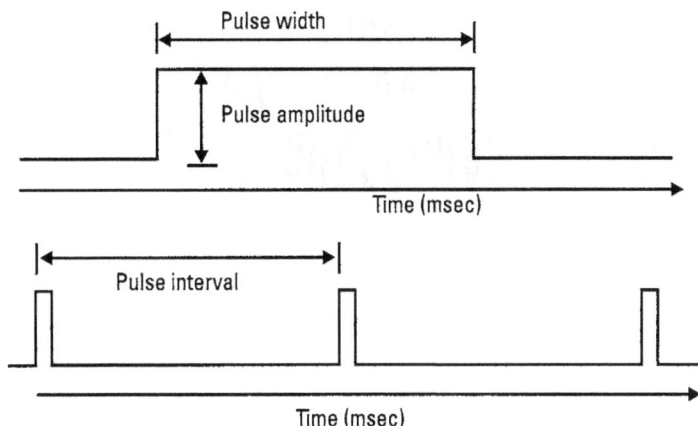

Figure 2.3 Pulse-modulation parameters.

- Because pulses are quite short, the time between transmission and reception of signals can be easily measured.

- The sharing of a single antenna by a transmitter and a receiver is simplified because the high transmitted power is isolated (in time) from the very low power of the received reflections.

- The average power that must be supplied to a radar transmitter is much less than the peak power, which makes both transmission tubes and power supplies smaller and lighter.

As shown in Figure 2.4, the actual pulse transmission is an RF signal turned on for the duration of the pulse. The video pulse (shown in Figure 2.3) is the signal used to modulate the pulse transmitter to create the RF pulse. It is also the signal output by a receiver when it detects a pulse.

Figure 2.5 shows the block diagram of a pulsed radar. Note that the transmitter and receiver share a common antenna. Since the pulse has a short duty cycle, the duplexer keeps the antenna connected to the receiver most of the time. Since the transmitter power is usually high enough to damage the receiver, the duplexer must have enough isolation to protect the receiver during the transmitted pulse.

The modulator turns the transmitter on during the pulse. The receiver receives and demodulates the reflected pulse. The display and control accepts mode commands from an operator and displays return signals on one of several types of screens to indicate the location of targets. Note that this location

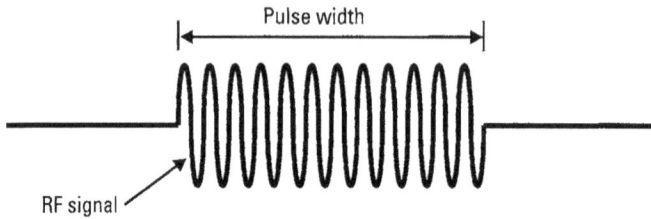

Figure 2.4 Transmitted pulse signal.

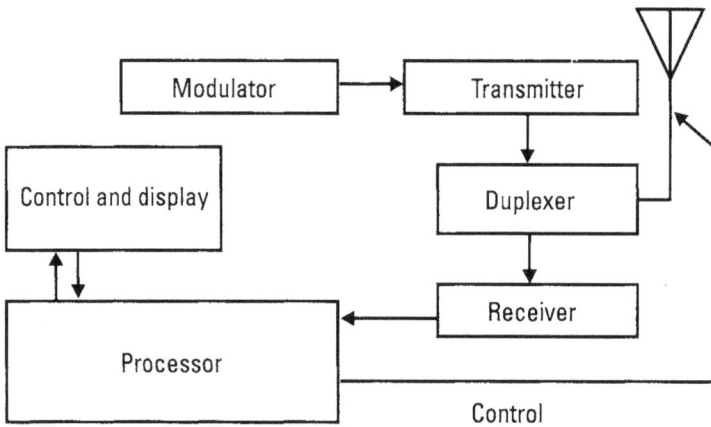

Figure 2.5 Block diagram of pulsed radar.

information can also be used to control missiles and the like. The processor determines the round-trip delay of reflected pulses, controls the antenna pointing, and prepares data for display or handoff of target information.

Continuous-Wave Radar

A continuous-wave (CW) radar requires two antennas, as shown in Figure 2.6, since the transmitter has 100% duty cycle. Isolation between these two antennas must be great enough to prevent the transmitter from saturating the receiver.

The CW radar compares the frequencies of the transmitted and received signals to determine relative velocity of the target from the Doppler shift. This is valuable in separating target returns from ground returns (i.e., radar reflections from the ground) in airborne radars with look-down-shoot-down capability.

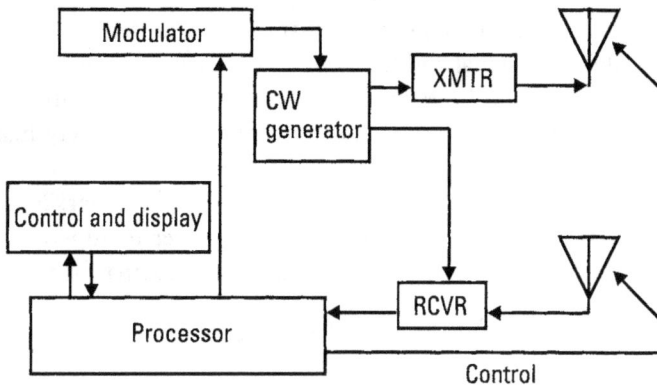

Figure 2.6 Block diagram of CW radar.

Frequency modulation waveforms such as those in Figure 2.7 allow a CW radar to determine the distance to a target as well as the relative velocity. Since the linear ramp portion of the waveform has a known time rate of change of frequency, the frequency difference between the transmitted and received signals is a function of both the Doppler shift and the round-trip distance to the target. The Doppler shift is determined during the flat part of the waveform both to isolate the range component and to provide velocity information.

Pulse Doppler Radar

Another type of radar that combines some of the features of both of the above types is the pulse Doppler radar. It differs from standard pulse radar in

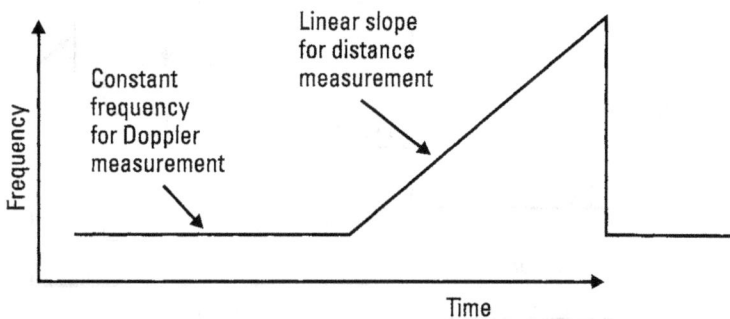

Figure 2.7 Modulating waveform for CW radar.

that its pulses are coherent. This means that the transmitted pulses are a continuation of the same signal and have phase consistency. Thus, the radar can detect reflected pulses coherently—giving the receiver a sensitivity advantage. High-PRF pulse Doppler radars are characterized by very high pulse repetition frequency and high duty cycle (30% to 50%).

Figure 2.8 is a block diagram of a pulse Doppler radar. Note that a sample of the CW generator is passed to the receiver to support coherent detection. The duplexer allows the use of a single antenna for transmission and reception.

2.1.3 Radar Cross Section

The amount of energy reflected to the radar receiver from a target depends on the radar cross section (RCS) of the target. In radar equations, the RCS is usually represented by the symbol σ. RCS is typically defined for monostatic radars, but for bistatic radars there is also a bistatic RCS. RCS is a function of the geometric cross section of the target, its reflectivity, and its directivity.

The geometric cross section is the size of the target as viewed from the aspect of the radar. The reflectivity is the ratio of the power leaving the target versus the radar power illuminating the target. The rest of the power is absorbed. Directivity is the ratio of the power scattered in the direction of the radar receiver versus the amount of power that would have been reflected if the total power were scattered equally in all directions. If the radar is monostatic, this is for the power scattered back toward the radar transmitter

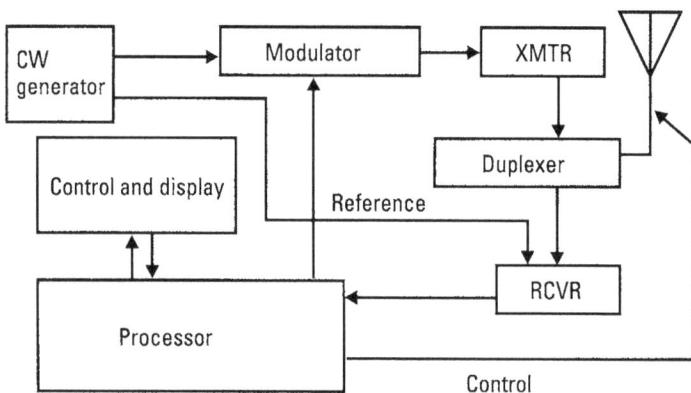

Figure 2.8 Block diagram of pulse Doppler radar.

(also the receiver location). If it is a bistatic radar, this is for the energy scattered from the transmitter location toward the receiver location.

RCS for a specific target is typically shown as a function of azimuth around the target. Figure 2.9 shows the RCS versus azimuth for a typical (older type) tactical aircraft. This would be the RCS for a specific radar frequency and assumes that the radar is in the aircraft's yaw plane. The units dBsm will be defined in Chapter 3. RCS charts like this are generated for individual, tactical aircraft at several frequencies. The charts are usually classified, since they reveal the vulnerability of aircraft to radar detection, but can be acquired with proper authorization. The character of the RCS pattern is caused by features of the aircraft. From the front and rear, the radar can see into the engines. At higher radar frequencies, the engine parts reflect significant energy. From the sides of the aircraft, the radar signal is efficiently reflected by the larger cross section of the fuselage and angles between the wings and fuselage. Note that both of these effects are reduced in modern aircraft designed for reduced RCS.

The RCS for a ship is described in similar charts. Figure 2.10 shows the RCS for an older destroyer. It is for a single radar frequency and assumes that the radar is at an elevation angle above the water at which the RCS is maximum. Note that there are very high peak RCS values 90 degrees from the bow and lesser peaks fore and aft. The RCS at the rest of the angles around the ship have been aptly described as a fuzz ball because of the many narrow peaks caused by individual reflective features of the ship. Modern ships designed for reduced RCS have significantly lower reflections, but large ships still have significant RCS.

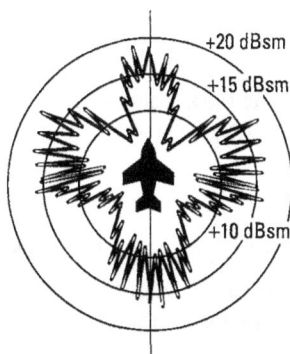

Figure 2.9 Typical tactical aircraft RCS.

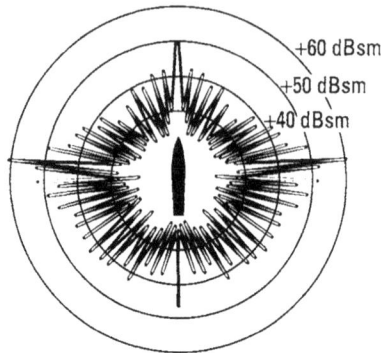

Figure 2.10 Typical ship RCS.

2.1.4 Radar Performance

Radar performance is described in terms of several important equations, listed below, that will be discussed in detail in Chapter 4.

- The radar range equation yields the signal level received by a radar receiver as a function of the transmitter power, the antenna gain, the range to the target, the operating frequency, and the RCS of the target.

- The detection range of the radar is the range at which the radar can achieve sufficient signal in its receiver to adequately measure the range and angle(s) to a target of a specific RCS level. This equation is derived from the radar range equation.

- The detectability range of the radar is the range at which a hostile receiver can detect the radar's signal. This must be defined in terms of the performance specifications of the receiver.

A radar resolution cell is the space within which a radar cannot determine whether it sees a single target or multiple targets. Figure 2.11 shows a typical radar resolution cell in two dimensions. The width of the cell is the antenna's beam width, and the depth of the cell is one half of the radar's processed pulse duration—both converted to units of distance. The equation for the width (w) of the resolution cell is

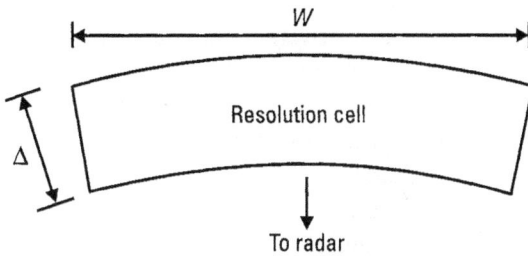

Figure 2.11 Radar resolution cell.

$$w = 2\sin(0.5 \text{ beamwidth})$$
$$\times \text{distance from radar to target}$$

The equation for the depth (Δ) of the resolution cell is

$$\Delta = 0.5 \text{ pulse width} \times \text{speed of light}$$

As the distance between a radar and its target decreases, the width of the resolution cell decreases, but the range resolution stays the same.

Minimum and Maximum Unambiguous Range

The minimum range is determined by the pulse width. Basically, the pulse must end before its leading edge is reflected from the minimum range target and returned to the receiver.

The maximum unambiguous range of a radar is determined by its pulse repetition interval. To avoid ambiguity, the reflection of one pulse from a target at maximum range must be received before a second pulse is transmitted. Otherwise, the radar would not be able to determine which pulse delay represented the actual range to the target.

2.1.5 Types of Radars

We will classify radars in terms of the tasks that they perform. Each radar type has characteristics that strongly affect the way it interacts with EW systems, and hence the way they are represented in EW modeling and simulation.

Search Radars

Air search radars are sometimes called EW/GCI radars because of their roles in early warning and ground control intercept. These radars typically have very long range. They are at relatively low frequencies and have long pulse duration. Some search radars have *chirped* pulses (i.e., pulses with linear frequency modulation). This allows the range resolution of the radar to be improved. Range resolution can also be improved through the use of binary codes during the pulse. Search radar antennas are usually large, providing relatively narrow antenna beams.

When a search radar acquires a target, the radar hands it off to a radar associated with a weapon system or passes information to a fighter aircraft controller who guides a fighter to intercept target aircraft that are detected. Sometimes, specific search radars are associated with specific weapon systems. Also, a single radar can have both acquisition and tracking modes.

Tracking Radars

Tracking radars provide sufficiently high location accuracy and frequent location updates to allow weapons to engage their targets. Radar guided missiles are actually guided to the constantly updated locations of their targets. Guns require three-dimensional tracking information so that they can be pointed and their shell fuses set to the appropriate delay to cause the fired rounds to explode at the target location.

Tracking radars typically operate at higher frequencies than search radars. They also have shorter pulses and higher pulse repetition rates. They are designed to operate over a little more than the lethal range of the weapons they support.

Battlefield Surveillance Radars

Battlefield surveillance radars are Doppler radars that can detect and locate vehicles and personnel moving over the ground, day or night.

Synthetic aperture radars are airborne radars that map terrain. The term *synthetic aperture* refers to the fact that data is combined over a period of time as the aircraft moves. This creates the effect of a very large antenna and the resulting resolution.

Moving target indicators are Doppler radars that detect moving targets over a large area. They divide the area of coverage into cells and determine the cells that contain moving targets.

Low-probability-of-intercept (LPI) radars are characterized by detectability range less than detection range. They use complex waveforms and careful parametric trade-offs to achieve this result.

2.1.6 Missile Guidance Techniques

There are seven techniques used to guide a missile to a target: command, active, semiactive, passive, imaging, terrain following, and GPS. Each has typical applications, and each is subject to EW techniques. Some missile systems use different techniques during different portions of their target engagement.

Command Guidance

Command guidance is common in surface-to-air missiles. As shown in Figure 2.12, the radar tracks the target and predicts its future locations. Guidance commands are generated and sent to the missile, taking into consideration moment-to-moment changes in the path of the target during the engagement. With command guidance, the radar has all of the information; the missile simply goes where it is told.

Active Guidance

An actively guided missile has a complete radar aboard, as shown in Figure 2.13. This type of missile typically turns on its radar as soon as it is within range of the target. Then it acquires the target and guides the missile to a meeting with the target. This is the most common type of antiship missile guidance, and is increasingly common in antiaircraft missile systems.

Semiactive Guidance

A semiactively guided missile has only a radar receiver on board (see Figure 2.14). The transmitter is remote from the missile. The transmitter tracks the target to keep its illuminator in place, or uses a very wide transmit antenna beam. The receiver receives the reflected signals from the target illuminator and guides itself to intercept the target. This is an example of a bistatic radar, since the transmitter and receiver are separated. It is commonly used in air-to-air missiles.

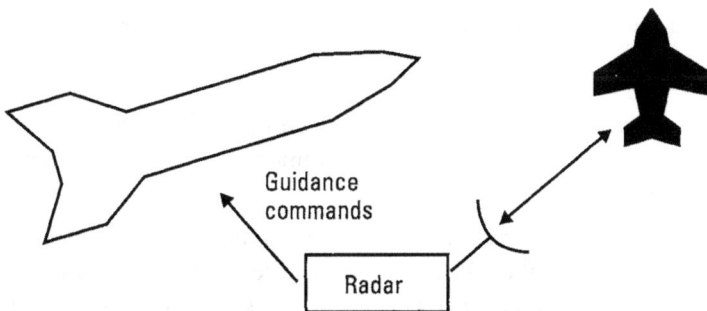

Figure 2.12 Command guidance.

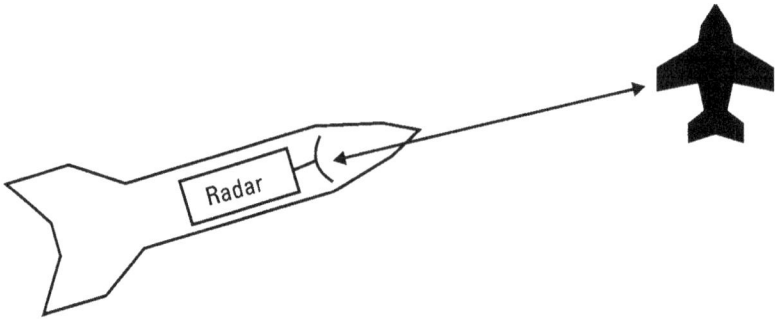

Figure 2.13 Active guidance.

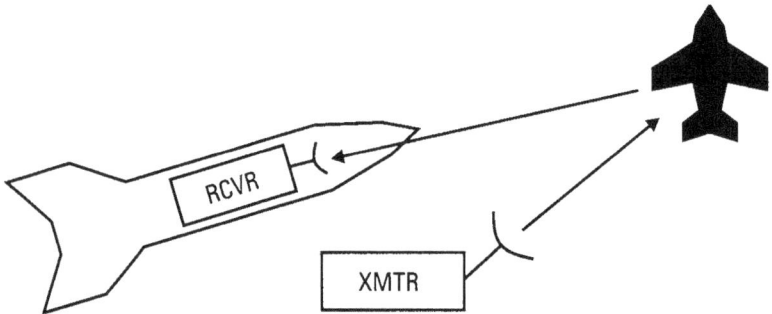

Figure 2.14 Semiactive guidance.

Passive Guidance

A passively guided missile homes on some energy emanating from the target, as shown in Figure 2.15. This can be radio frequency energy or infrared (IR) energy. Three important examples of this type of guidance include the following:

1. *Home on jam modes in surface-to-air missiles.* If the enemy determines that a target is using a jammer, the missile guidance will go to a passive mode to home on the jammer transmission.

2. *Antiradiation missiles.* These are set to the parameters of a particular transmitter; they then guide themselves to the transmitter location.

3. *Heat-seeking missiles.* These missiles have IR trackers and guide themselves toward the source of the heat (for example, the engine of a jet aircraft).

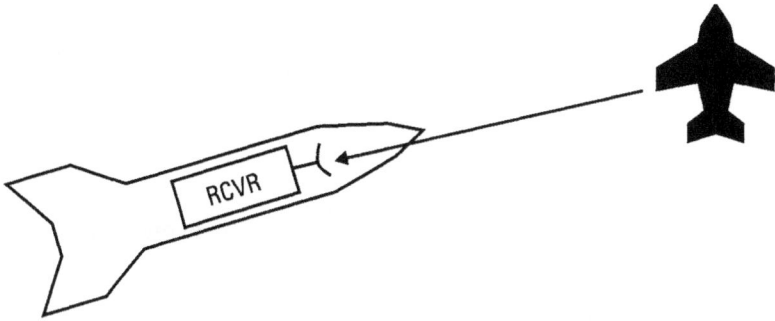

Figure 2.15 Passive guidance.

Imaging Guidance

An imaging-guided missile has an imaging sensor that captures an image of a target and centers that image in its sensor field of view to guide the missile to the target, as shown in Figure 2.16.

Terrain-Following Guidance

When a missile matches the terrain below it with a map of the terrain over which it is programmed to fly, the missile is said to use terrain-following guidance. This is normally associated with a radar and a radar map, as shown in Figure 2.17, but can also follow any other type of terrain definition.

GPS Guidance

If the exact latitude and longitude of a target is known, a missile can fly to that location using a GPS receiver to determine its own location (see Figure 2.18).

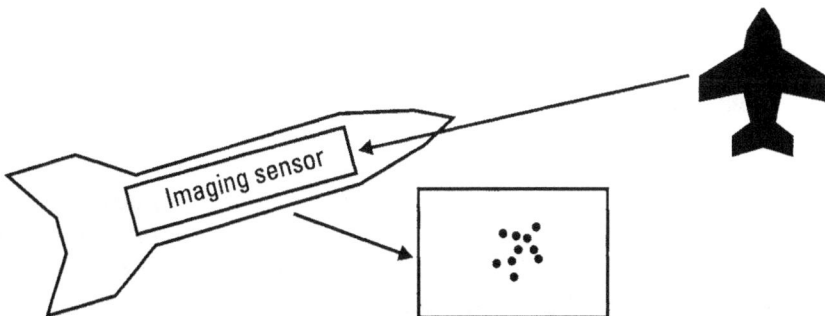

Figure 2.16 Imaging guidance.

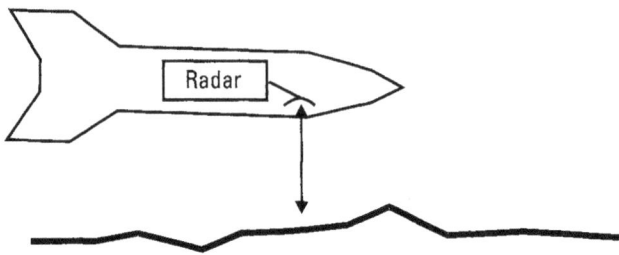

Figure 2.17 Terrain-following guidance.

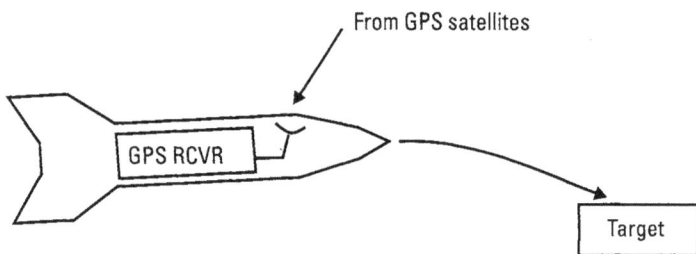

Figure 2.18 GPS guidance.

2.1.7   References for Further Study

An excellent text on radar principles and operation for those with little or no background in the field is *Introduction to Airborne Radar,* by George W. Stimson (2d ed., Medham, NJ: SciTech, 1998). It is available from the Association of Old Crows (www.crows.org). For a more technical coverage of the subject, see *Introduction to Radar Systems,* by Merrill Skolnik (New York: McGraw-Hill, 1962).

2.2   Communication

Communication signals are also important to EW considerations. They are intercepted and jammed, and their externals are analyzed to determine an enemy's electronic order of battle. The messages carried by a communication system are called the signal *internals,* while the modulation, transmitter location, and so forth are called its *externals.*

While radar signals are typically transmitted from narrow-beam antennas, communication signals are usually transmitted from antennas that cover

360 degrees of azimuth. Also, since communication signals are designed to carry information, they tend to have high duty factor (with notable exceptions, such as burst transmissions) and to use continuous type modulations.

As shown in Figure 2.19, communication involves one-way links. Each link includes a transmitter, a transmitting antenna, a receiving antenna, a receiver, and everything that happens to the signal in between. The purpose of the communication link is to get information from the location of the transmitter to the location of the receiver. Equations to predict the performance of one-way links will be covered in Chapter 4.

2.2.1 Tactical Communications

The most significant type of communications signals for EW consideration support tactical communication associated with military organizations. Tactical radios are normally transceivers (i.e., they transmit and receive through one antenna). They are typically organized into nets, as shown in Figure 2.20. All of the radios in a net typically operate on a single frequency, so only one member of the net can transmit at a time. In a military organization, the

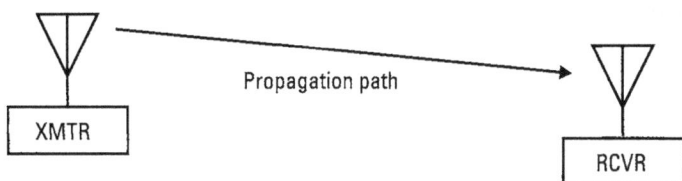

Figure 2.19 Basic communication link.

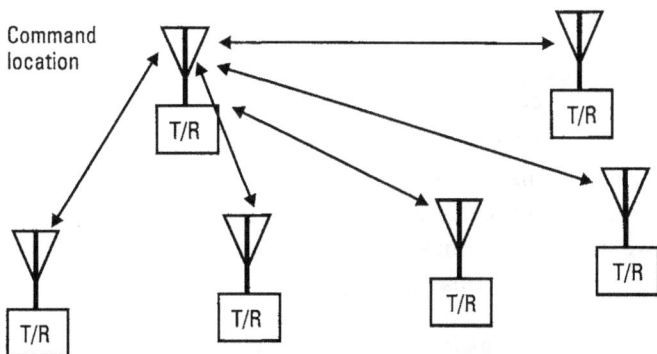

Figure 2.20 Tactical communications net.

communication normally goes primarily from the commander to the subordinate units. Although all members of the net can talk to each other, the normal mode of operation is for the command transmitter to transmit longer messages—more often than the subordinate transmitters. The subordinate members of the organization tend to make short responses to transmissions from the commander.

In any type of modern military situation, a great deal of radio communication is required. Therefore, the available frequency spectrum is heavily used. A common assumption for military communication bands is that 5% to 10% of the spectrum will be occupied at any instant. However, if you observe the spectrum over a few seconds, every available channel will typically be full.

The duration of tactical communications transmissions can be expected to be as short as one second. Under extreme communication discipline, replies to command messages may even be as short as a key click (less than half a second). Tactical communication includes air-to-air, air-to-ground, and ground-to-ground communications. Each has assigned frequency bands.

2.2.2   Data Links

Another type of communication signal is the data link. Links are used to pass command and data information between unmanned aerial vehicles (UAV) and their ground stations, between aircraft systems that share information, and between satellites and their ground stations.

Data links are actually called different names, as shown in Figure 2.21. There are command links (or uplinks) that send command signals to UAVs. True data links, or downlinks, send data (typically wideband) from sensors on the UAVs to their ground stations. There are also telemetry links (narrowband) that carry status information about the UAV and payloads to the ground stations. Another type of link is a broadcast link that passes information from sensors onboard the UAV to multiple authorized receivers at ground locations. Satellites with military sensors, manned aircraft carrying remotely controlled EW assets, and remotely controlled ground-based EW systems have the same types of links.

Communication satellites (and manned or unmanned aircraft acting as relays) have wideband uplinks and downlinks as shown in Figure 2.22. Both links carry the same information in most cases.

2.2.3   Communication Bands

Military communication is in the high-frequency (HF), very-high-frequency (VHF), ultra-high-frequency (UHF), or microwave-frequency ranges. There

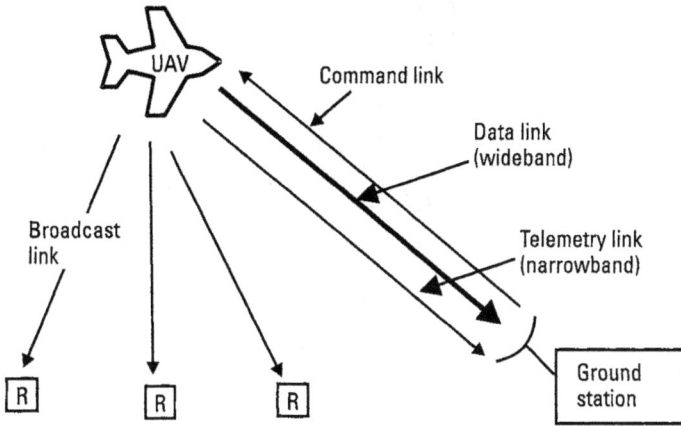

Figure 2.21 Links supporting a UAV.

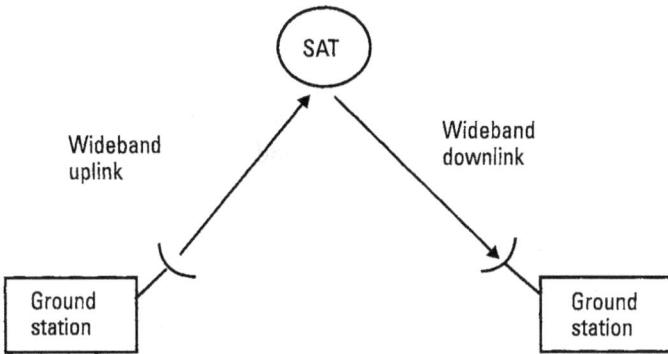

Figure 2.22 Communication satellite links.

is also some communication in lower frequency ranges and at IR or laser frequencies (see Figure 2.23). Each of these frequency ranges has different
propagation characteristics.

In general, the higher the frequency range, the more information the
link can carry. Another generality is the higher the frequency, the more link
performance is degraded by non-line-of-sight conditions.

Low-Frequency Ranges

At the lowest-frequency ranges, including very low frequency (VLF), only a
few hertz of information can be transmitted. But, on the other hand, it can

Frequency
(Hz)

10^{15} — Optical

10^{14}

10^{13} — IR

10^{12}

10^{11}

10^{10} — Microwave

10^{9}

10^{8} — UHF

    — VHF

10^{7}

    — HF

10^{6}

    — LF and VLF

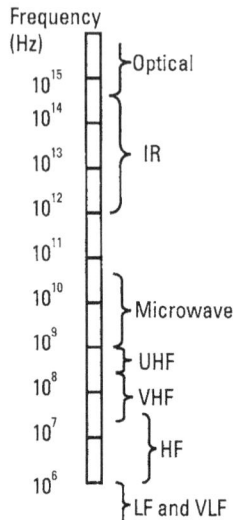

Figure 2.23 Frequency bands for communication.

be transmitted clear around the world. If you want to communication with submerged submarines, VLF is the right band.

HF

The HF band is 3 to 30 MHz. It supports wide-enough channel bandwidths to carry voice information, and does not require line of sight. HF signals are reflected by the ionosphere, causing hops that can propagate the signals clear around the Earth. HF signals also have significant ground-wave propagation well beyond the line-of-sight horizon.

VHF

The VHF band is 30 to 300 MHz, but receivers and transceivers called VHF often operate from about 20 to 250 MHz. They can carry voice and video information. (U.S. television channels 2 through 13 are in the VHF range.) Although VHF signals operate primarily line of sight, they can propagate around the curve of the Earth and over ridgelines with an additional attenuation that is a function of the frequency and the geometry of the transmission path.

UHF

Although the UHF band is 300 to 3,000 MHz, UHF receivers and transceivers most often operate from about 250 MHz to 1 GHz. They carry voice, video, and digital data. These signals are also considered line of sight—extended by the same non-line-of-sight modes as VHF signals. However, since these modes are all frequency dependent, the UHF signals have significantly more non-line-of-sight attenuation than VHF signals in the same transmission geometry.

Microwave

Microwave communication signals (1 to about 40 GHz) are considered completely line-of-sight, but they can carry a great deal of information. Also, because of the much greater attenuation of these signals with distance (see Chapter 4), they usually require directional antennas. This means that wideband communication links between fixed locations are very practical, but communication involving moving platforms requires complexity in the antenna implementation.

Signals in the high microwave range are also significantly affected by atmospheric attenuation. In normal applications, signals above 10 GHz are also significantly attenuated by heavy rain. At lower microwave frequencies there is moderate rain loss, and at UHF and below, rain loss is typically ignored.

Data links are normally implemented in the microwave range. Ground stations and most airborne platforms have directional antennas that must be steered to point at each other. Likewise, satellites and their ground stations require steerable antennas. The exception to this is the synchronous satellite, which remains in a fixed position over the Earth. Ground antennas can be left in a fixed orientation, and satellite antennas either cover the whole Earth or point at ground stations that remain in fixed geometry relative to the satellite.

IR and Optical Ranges

Communication in the IR and optical frequency ranges allows the transmission of vast amounts of information, but is restricted to the optical line of sight. It is severely degraded by fog or smoke.

Radio Line of Sight

It should be noted that the line of sight used for VHF, UHF, and microwave communication links—as well as for radars—is limited by the radio horizon rather than the geometric horizon caused by the curvature of the Earth. A maximum length transmission path over smooth Earth is shown in Figure 2.24.

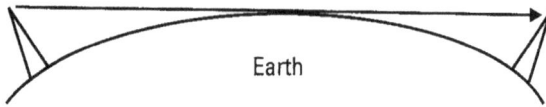

Figure 2.24 Maximum-length transmission path.

The radio horizon is farther away than the geometric horizon because radio waves are refracted in the Earth's atmosphere. While this refraction can vary widely, there is an approximation that is almost universally accepted.

This approximation is commonly called the *four-thirds Earth effect*. If the Earth were one-third larger in diameter, geometric (or optical) line of sight over the curved Earth would reach to a greater distance because a larger sphere has less curvature. An equation for range limitation with four-thirds Earth is presented in Section 6.5.

2.2.4 Communication Modulations

The most common modulations used for communication are amplitude modulation (AM), frequency modulation (FM), phase modulation (PM), and single sideband (SSB). Modulations for digital signals and new LPI modulations, although they use these basic modulations, warrant special discussion.

Each of these modulations carries the communicated information. Each modifies a carrier signal at the frequency of operation of the link.

AM

Amplitude modulation changes the amplitude of the carrier signal in accordance with the information that the signal communicates. A radar pulse is a special case of AM. Video signals for broadcast television are amplitude modulated. The bandwidth of AM signals is directly proportional to the amount of information they carry. In general, the transmission bandwidth is twice the bandwidth of the information carried. Figure 2.25 shows the AM spectrum (i.e., power versus frequency). Note that the upper and lower modulation sidebands are mirror image, with each carrying all of the information transmitted.

FM

Frequency modulation changes the frequency of the carrier signal in accordance with the modulating information. One characteristic of FM is that the transmitted frequency range over which the carrier is changed can be arbitrarily set. The wider the FM range, the more resistant the signal becomes to

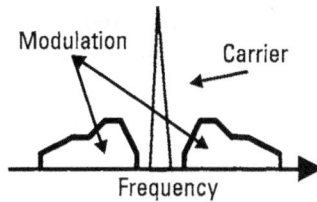

Figure 2.25 AM signal spectrum.

interference. The ratio between the transmission bandwidth (the FM range) and the amount of information carried by the signal is defined by the modulation index (β).

SSB

Single sideband signals are basically AM signals with the carrier and one of the sidebands filtered off. This makes the SSB transceivers more complex than AM transceivers, but allows for more efficient use of frequency spectrum. SSB signals can be either upper sideband (USB) or lower sideband (LSB), depending on which AM sideband is transmitted.

PM

In phase modulation, the phase of the carrier signal is changed in accordance with the modulating information. For narrowband communication, PM is very similar to FM, but when the signal carries digital information, the picture is more complex.

Digital Communication

When a signal carries digital data, the data itself is ones and zeros, called *bits*. However, these bits must be converted to a form that can be imposed on a carrier signal before they can be transmitted. Common modulations include the following:

- Amplitude shift keying (ASK), where zero and one are different levels of amplitude modulation;
- On-off keying (OOK), where a pulse is transmitted for a one and a space for a zero;
- Frequency shift keying (FSK), where zero and one are different frequencies;

- Phase shift keying (PSK), where zero and one are different phases. PSK can be binary (0- or 180-degree phase shift) or can use many phase positions to carry more data in less bandwidth.

LPI Communications

LPI communication is designed to reduce the ability of hostile listeners to detect the presence of signals, to intercept them, to jam them, and to locate transmitters. To achieve this, transmitters use one of three types of special modulation: frequency hop, chirp, or direct sequence. These are three spread spectrum techniques that reduce the signal power at any one frequency. A receiver in the same net will have a synchronization scheme that allows it to reconstruct the original information signal sent, as shown in Figure 2.26. However, a hostile receiver will not be able to despread the signal and thus must deal with a much lower signal strength (usually by a factor of thousands). In addition, a signal of the same type that is not spread by the synchronization scheme will be (in effect) spread by the despreading process, so the synchronized receiver will ignore it.

Frequency hop involves changing the transmission frequency 100 or more times per second. The transmitter hops randomly over a frequency range of many times the transmission bandwidth. A synchronization scheme allows receivers in the same net to hop along with the transmitter. Hostile receivers have no way to synchronize with the transmitter, so they are at a great disadvantage when trying to detect or copy the signal. Since the signal

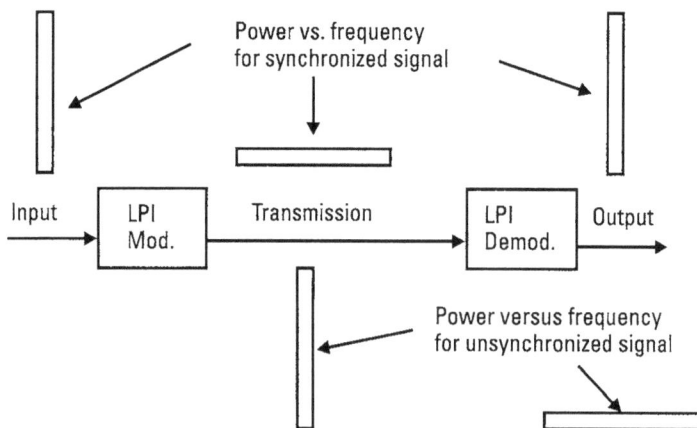

Figure 2.26 LPI signal performance.

moves quickly over a wide frequency range, a jammer must either cover the whole range or find some way to measure the frequency each time the signal hops to tune a jammer.

Chirp signals sweep very rapidly across a frequency range much wider than the transmission bandwidth. Random sweep intervals or non-linear sweep patterns prevent a hostile receiver from synchronizing to the sweep for detection or jammer control.

Direct-sequence signals have a second modulation applied to spread the frequency of the transmitted signal. This modulation is a high-rate, pseudorandom bit stream that usually PSK modulates the information signal. The receiver is synchronized to the same bit stream, so it can remove the spreading modulation. However, a hostile receiver cannot despread the signal and cannot typically even detect the signal's presence. Jamming signals are spread by the same demodulation that despreads the desired signal, making jamming very difficult.

2.3 Electronic Support

Electronic support (ES) was previously called *electromagnetic support measures* (ESM) and is still often called by that name. It is the listening part of EW. ES systems and subsystems detect threat signals and display the types and location of transmitters to support situation awareness or cue electronic attack capabilities.

The primary ES functions are detection of threat signals, identification of threat types and operating modes, location of threat emitters, specific emitter identification (SEI), and display or handoff of threat information. The most important ES system types are radar warning receivers (RWRs), threat-targeting systems (often called *ESM systems*), and battlefield surveillance systems.

ES systems operate in both the radar and communication bands, providing similar functions. However, the nature of the signals dictates different approaches. There are also differences in the way tasks are approached for airborne, ground-based, and shipboard applications.

2.3.1 Radar Warning Receivers

The purpose of an RWR is to identify the presence of threats to an aircraft quickly so that countermeasures or maneuvers can be taken to defeat the impending attack. An RWR is designed to operate in a high-density signal environment. It can typically accept millions of pulses per second along with

CW radars. It is usually required to identify and report new threats within approximately one second.

The RWR, by identifying threat radars, determines each actual threat that is present. This will be a particular type of surface-to-air missile (SAM), Radar-controlled antiaircraft gun (AAA), or fighter aircraft with aerial intercept (AI) radar. By analysis of the radar's parameters, the RWR also determines the location of the threat relative to the protected aircraft and the operating mode of the threat. Threat operating modes are usually search, tracking, or launch.

Figure 2.27 is a block diagram of a basic RWR. The four antennas are located somewhat symmetrically around the aircraft, as shown (not to scale) in Figure 2.28. They are usually mounted about 15 degrees depressed from the aircraft's yaw plane. Together, they give instantaneous 360-degree azimuth coverage from a few degrees above the yaw plane to more than 45 degrees below the yaw plane. The four receivers located with the antennas almost always include multiple crystal video receivers, to receive pulses over a very-wide-frequency range. In modern RWRs, they may also include narrowband receivers to handle CW and pulse-Doppler radar signals. These receivers usually have filters to divide the frequency range into bands, and may also include band stop filters to remove high duty rate signals from the inputs to the crystal video receivers. In Section 5.2, we will discuss these different types of receivers in detail, along with their performance parameters and the ways their performance is simulated.

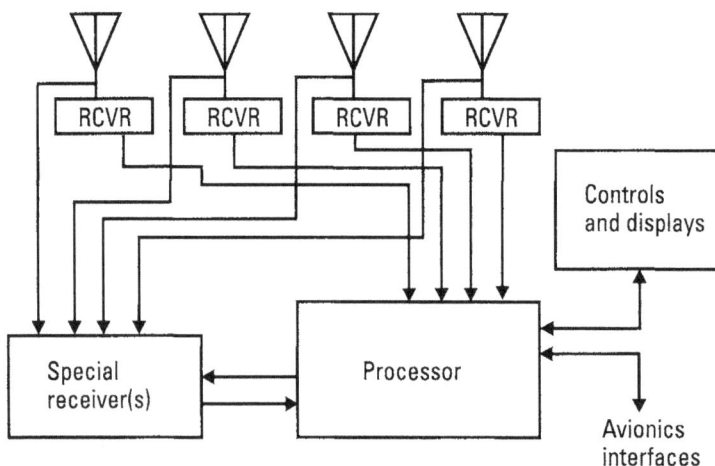

Figure 2.27 Block diagram of typical RWR.

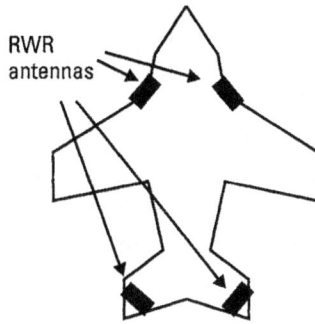

Figure 2.28 Typical RWR antenna installation locations.

The special receiver can be one or more instantaneous frequency measurement (IFM) receivers to measure signal frequency on a pulse-by-pulse basis. It can also include channelized or digital receivers to handle difficult signal conditions.

The processor controls all of the other components in the RWR system, identifies threat types and modes, and determines emitter locations. It also provides outputs to cockpit displays and sometimes headsets, accepts control commands from the aircrew, accepts coordinating and handoff inputs from other avionics systems, and formats outputs to countermeasures systems. The processor can be a single computer, or several computers to handle individual processing tasks.

RWRs typically have monopulse direction-finding capability. That is, they measure the direction of arrival (DOA) of every pulse received. Older RWRs compared the received pulse amplitude at multiple antennas to determine DOA. Some modern systems achieve more accurate DOA with a phase comparison direction-finding (DF) scheme. These two approaches and several others will be described in detail in Section 5.5.

RWR processing starts with the separation of signals. Since most RWRs use wideband pulse receivers, they receive interleaved pulses from many pulse trains. Each pulse train is from a single threat emitter. If the RWR can measure the RF frequency of each pulse, this can be a powerful differentiating parameter. Otherwise, the RWR can use algorithms that look for patterns in the pulses, to separate inputs by pulse intervals of known threat types. In any case, the direction of arrival of pulses can be a powerful tool to assist in the separation of individual threat signals.

Once individual signals are isolated, they must be identified. This is done by comparing the measured signal parameters against a threat identification table (TID). Parameters collected and evaluated include RF frequency, pulse width, pulse-repetition frequency, modulation (chirp or binary code) on pulses, CW modulations, and antenna scan characteristics. When the measured parameters for a threat signal match one of the sets of parameters in the TID, the presence of the corresponding threat is declared. If the collected parameters do not match any parameter sets in the TID, the signal type is declared to be unknown. In some modern systems, the computer goes on to determine the closest fit to force an identification.

RWR displays have classically involved a three-inch round tube in the cockpit that displays the location of threats. The first such displays showed strobes on the screen in the direction of the threat emitter—with the nose of the aircraft represented by the top of the screen. The type of threat was shown by coding the strobes (e.g., dashed lines) and in a so-called billboard display that had small square buttons that lit up to show that a specific threat type was present. Later, the round-tube display just showed the threat type by a symbol, with the position of the symbol on the screen indicating the location of the threat. This type of display is currently used in many types of helicopters and fixed-wing aircraft. An example of this type of screen is shown in Figure 2.29.

As shown on the screen, the RWR indicates that an SA-3 SAM is some distance from the aircraft at 10 o'clock. An SA-2 is at 2 o'clock, an SA-6 (designated as the highest-priority threat) is at 3 o'clock, an AAA is fairly close at 9 o'clock, and an AI is dangerously close at 7 o'clock.

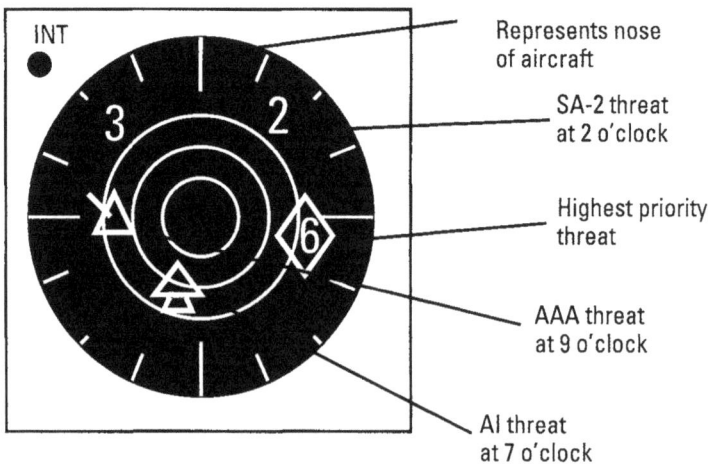

Figure 2.29 Typical computer-driven RWR display.

In the most modern tactical aircraft, the threats are shown as lethality zones on an integrated pictorial display. The RWR produces the same information, but it is presented to the pilot in a form that is integrated with other important information—such as terrain, planned flight-path, and so on.

2.3.2 Shipboard ESM Systems

Shipboard ES systems operate in an environment significantly different from that of the airborne RWR. Because of the different requirements, shipboard ES systems are usually called ESM systems. An ESM system typically has more sensitivity than an RWR, and takes more detailed data to support specific emitter identification (SEI) and targeting.

The horizon for a ship is only about 10 km, so its ES system cannot see many surface threats. It can see high-flying aircraft (there will only be a few in the blue water situation), and it can see surface radars within 10 km. (By contrast, the long horizon distance for an aircraft at 40,000 ft causes it to see millions of pulses per second.) However, the individual threats to the ship are extremely lethal because of the ship's limited mobility. Figure 2.30 shows a block diagram of a typical shipboard ESM system.

The four antennas feeding the wideband receiver are similar to those in the airborne RWR. Their purpose is to quickly acquire close-in threats. They are extremely important close to land (in the so-called *littoral* or *brown water* environment). These antennas are fixed around a mast to give 360-degree

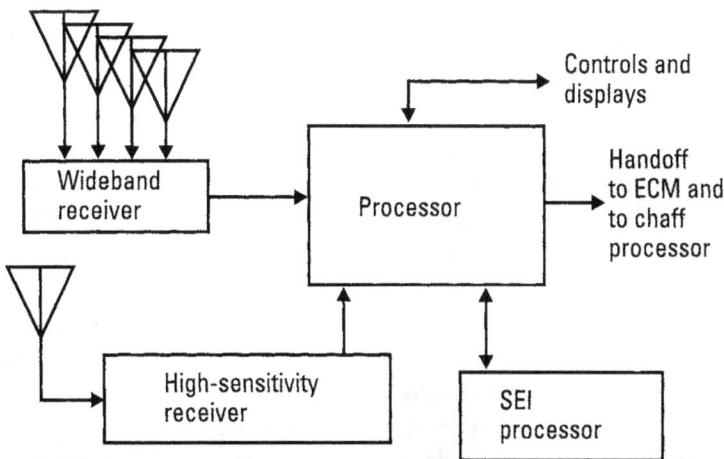

Figure 2.30 Block diagram of typical shipboard ESM system.

coverage. The wideband receiver is similar to the four pulse receivers on the RWR and may include an IFM and other types of receivers. The processor performs the same functions as the airborne RWR processor, but is customized to the appropriate threats. It hands off warning of a new threat within about one second.

The single antenna is normally a rotating dish that covers the whole threat band. The high-sensitivity receiver is typically a narrowband type that provides very high quality intercepts that allow the processing of complex modulations and specific emitter analysis.

The ESM system presents threat information to the crew so that evasive maneuvers can be initiated if appropriate. It also hands off threat type and azimuth of arrival to a chaff control processor that fires chaff rockets to lure antiship missiles away from the ship.

The SEI processor performs detailed analysis of such parameters as pulse rise time and frequency modulation on pulses to detect unintentional modulations that differentiate one ship's radar from that of another ship. When compared to extensive SEI tables, this allows a determination of the ship (by hull number) that the ESM system sees. People who make SEI systems claim that they can detect the presence of dents in the reflector of a radar antenna.

The displays of shipboard ESM systems provide detailed information about detected threats in alphanumeric as well as graphic formats.

2.3.3 Battlefield Surveillance Systems

Battlefield surveillance systems collect and analyze all of the radar and communications signals in an area of the battlefield to allow analysis of an enemy's electronic order of battle and to support targeting. These systems can be mounted in fixed-wing aircraft, helicopters, or ground mobile vehicles with erectable masts. In order to accurately determine the locations of emitters, multiple sensor platforms with highly accurate emitter locations systems are used simultaneously. Figure 2.31 shows the components of a typical fixed-wing aircraft system. The full system has three aircraft and a control-and-analysis center. There are no operators in the aircraft, but many operators on at the control and analysis center. As shown in Figure 2.32, direction-of-arrival measurements by three platforms cross at the emitter location. Although two platforms could triangulate to determine a location, the third line of bearing provides three location answers, allowing evaluation of the accuracy of the other two. If all three crossing points are very close, the location is given high confidence.

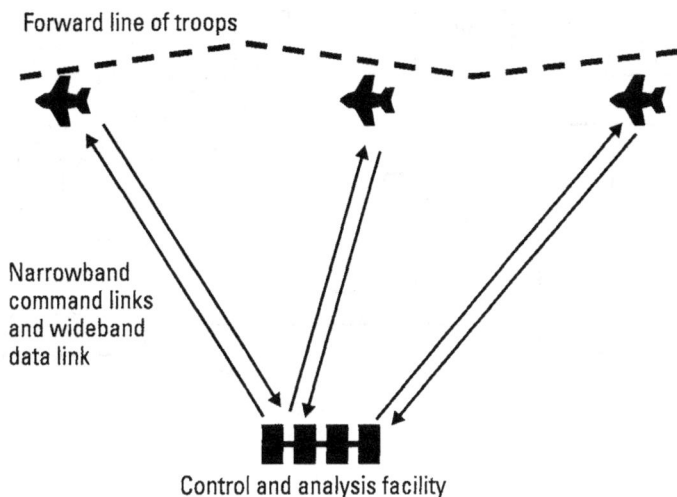

Figure 2.31 Battlefield surveillance system using fixed-wing aircraft.

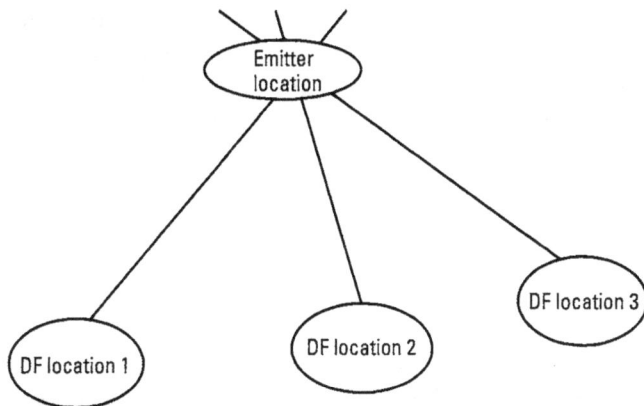

Figure 2.32 Triangulation to determine threat-emitter location.

Figure 2.33 shows a block diagram of the receiving and emitter-location system in a typical aircraft of the battlefield surveillance system. The multiple antennas are required to support the emitter-location capability. They also supply information for the individual monitoring receivers. There are many receivers, so that each operator can have exclusive use of a receiver for long-term analysis tasks. The emitter-location subsystem includes receivers that are tasked to support all operators as required. The precision

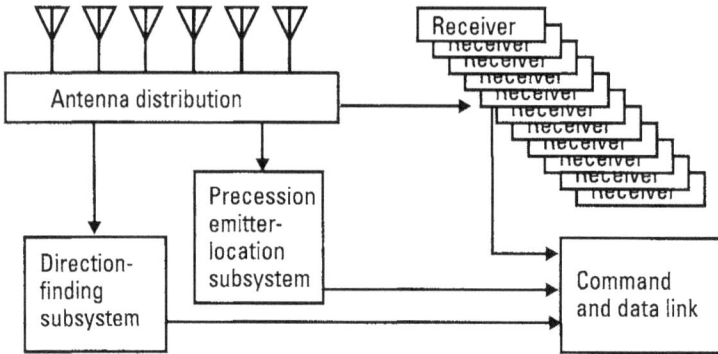

Figure 2.33 Block diagram of one aircraft's collection system.

emitter-location subsystem is tasked by operators who need to generate targeting data. Both the normal and the precision emitter-location techniques are described in Section 5.5.

The operators in the control-and-analysis center monitor and analyze signals received by the aircraft sensor systems, and when they request location information on a particular emitter, a special set of receivers in each of the three aircraft is tasked to take simultaneous angle-of-arrival readings.

Operator displays are as shown in Figure 2.34. Note that the map is a computer map available from the Defense Mapping Agency. The tactical situation information (organization locations, forward line of troops, etc.) is added locally. The system draws the locations of the platforms (1, 2, and 3) and the lines of bearing.

The output from the battlefield-surveillance control-and-analysis facility is a series of tactical situation reports to appropriate commanders and higher-level analysis centers. Appropriate tactical situation reports are also sent to lower-level commanders through a broadcast link from one of the system aircraft.

2.4   Electronic Attack

Electronic attack (EA) is the action part of EW. When ES finds out what is going on, EA does something about it. EA includes electronic countermeasures (ECM) and two other activities that were not previously considered part of EW. These are directed-energy weapons (DEW) and antiradiation missiles (ARM). We will first focus on jamming, then on the other portions of the field.

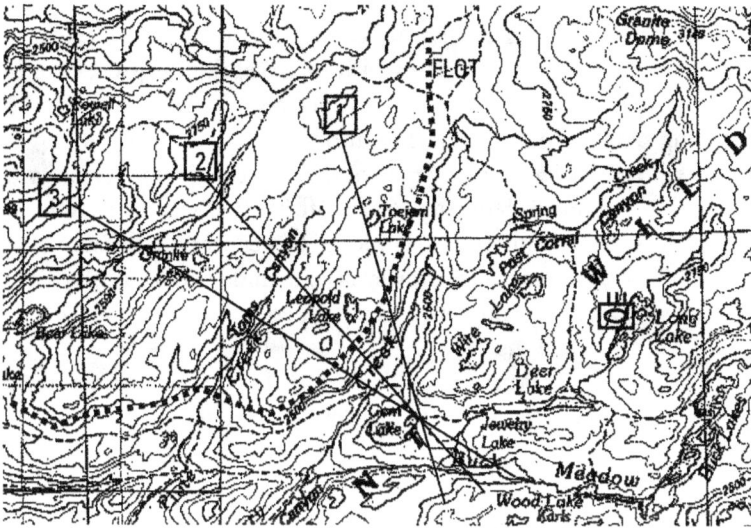

Figure 2.34 Typical operator display from battlefield-surveillance system. (Used with permission of Wildflower Productions).

2.4.1 Jamming

Figure 2.35 shows the basic jamming concept. Note that you always jam the receiver—not the transmitter. Jamming involves placing a signal into the receiver that interferes with the reception or processing of the desired signal. The confusion about jamming the transmitter comes from the fact that a monostatic radar has its transmitter and receiver in the same place.

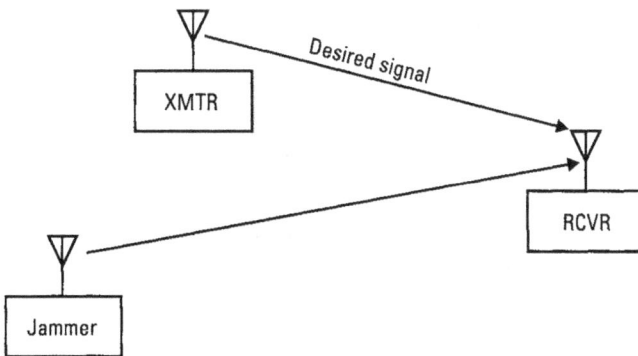

Figure 2.35 Basic communication-jamming geometry.

Figure 2.36 shows the radar-jamming concept. The jammer places a signal into the radar receiver that interferes with the reception or processing of the reflected signals returning from the target. For bistatic radar, you need to put the jamming signal into the receiver location.

We will classify jamming by type of threat signal (radar versus communication), by jamming geometry (standoff versus self-protection versus expendable), and by jamming technique (cover versus deceptive versus decoy). While it is somewhat controversial to include decoys as a type of jamming, they act a lot like jammers and their performance is calculated using some of the same equations.

Radar versus communication jamming is an easy distinction. Each attempts to interfere with the use of the electromagnetic spectrum for its corresponding enemy asset.

2.4.2   Communications Jamming

Communication jammers use the geometry that was shown in Figure 2.35. Their effectiveness is measured in terms of jamming-to-signal (J/S) ratio. Each signal (desired signal and jamming signal) propagates from its transmitter to the receiver. Each arrives at the receiver with received power that can be calculated from the formulas presented in Chapter 4. The comparison of the received power of the jamming signal divided by the received power of the desired signal is the J/S. The higher the J/S, the more effective the jamming. Unless the receiving antenna has some directivity (rare in tactical communication), the J/S is a function of the ratio of the jammer-radiated power to the desired signal-radiated power and the ratio of the square of the distance to the desired signal transmitter to the square of the distance to the

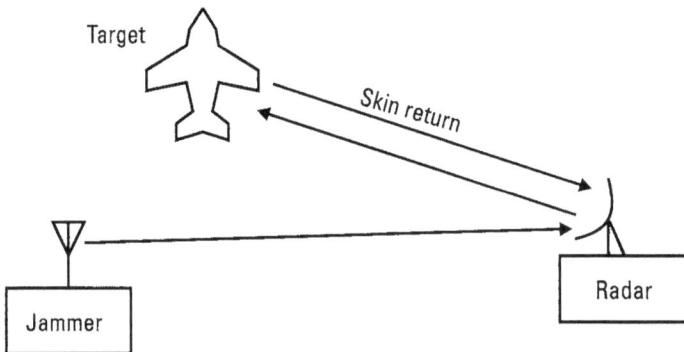

Figure 2.36   Radar-jamming geometry.

jammer. This is summarized in Figure 2.37. In general, a J/S of a factor of 10 is considered adequate for communications jamming.

When the communication link uses one of the LPI modulations, significantly more J/S is required to overcome the processing gain provided for the synchronized desired signal by the spreading modulation.

2.4.3 Radar Jamming

For radar jamming, consider the diagram shown in Figure 2.36. The J/S is the ratio of the power of the jammer that gets into the radar receiver to the power of the skin return in the radar receiver. There are several complicating factors to consider. First, we must consider the radar antenna directivity. The radar antenna is pointed at the target, so the skin return is increased twice by the gain of the antenna (once during transmission and once during reception). The transmitted jamming signal is increased by the jammer antenna gain (not the radar antenna gain) and is increased by receiving gain of the radar antenna only if the jammer is located on the target. Otherwise, the jammer must enter the radar antenna through its side lobes (see Section 5.1), which have much less antenna gain.

Another complicating factor is that the skin return to the radar is reduced by the fourth power of the distance from the radar to the target, while the jamming signal is only reduced by the square of the jammer to radar receiver distance.

Thus, as summarized in Figure 2.38, the J/S for radar jamming is a function of the fourth power of the distance to the target divided by the

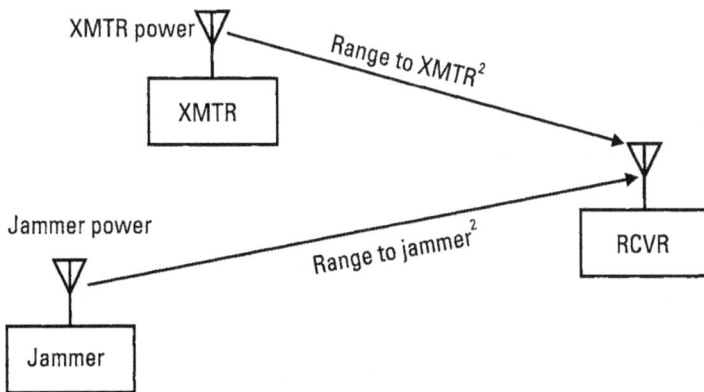

Figure 2.37 Communication-jamming J/S factors.

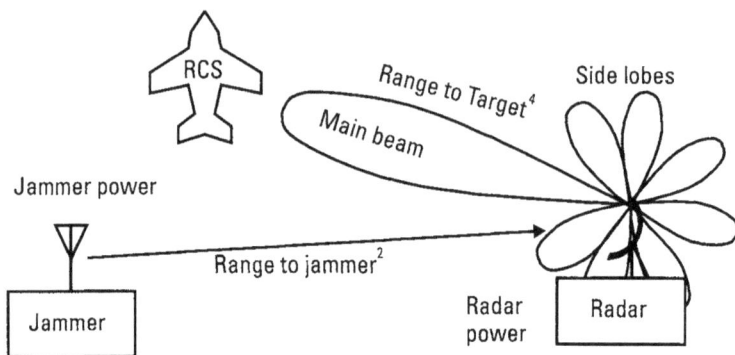

Figure 2.38 Radar-jamming J/S factors.

square of the distance to the jammer. It is also a function of the radiated jammer power to the radiated radar power and the ratio of the radar side lobe power to main beam power. The final factor is the RCS of the target. All else being the same, the smaller the RCS, the larger the J/S.

2.4.4   Standoff Jamming

Standoff jamming, as shown in Figure 2.39, is the protection of an attack aircraft inside the lethal range of an enemy weapon by a jamming aircraft that is

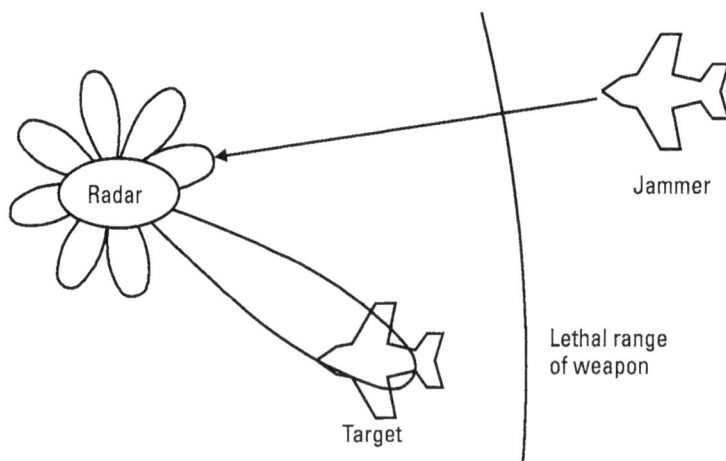

Figure 2.39 Standoff jamming geometry.

outside the lethal range of that weapon. The lethal range is the range at which the weapon can shoot down an aircraft.

Standoff jamming aircraft are large and carry large jammers with high radiated jamming power. Since they are farther away than the targets they are protecting, standoff jammers must have extra jamming power to overcome the square of the distance disadvantage. They must also have sufficient power to overcome the disadvantage of jamming into the radar antenna side lobes, since the radar is pointed at its target (i.e., the protected aircraft). The attacking aircraft normally have much less RCS, but they are closer to the radar. It should be noted that home-on jam modes of weapons are a significant threat to standoff jammers within the weapon's lethal range. When the attacking aircraft gets close enough to the radar that the radar can acquire a track, the radar is said to burn through the jamming. The radar-to-target range at which this occurs is called the *burn-through range*.

In general, a standoff jammer can prevent a radar from locking on to a protected aircraft, but does not have enough power to break the lock of a radar that is already tracking the protected aircraft.

2.4.5 Self-protection Jamming

In self-protection jamming, the jammer is, by definition, located on the target (see Figure 2.40). Thus the jammer has the advantage, being no farther from the radar than the target. It is also amplified by the radar's main beam gain. With these advantages, the self-protection jammer has sufficient power to break the lock of a radar that is tracking the target. However, if the enemy discovers that the target aircraft is jamming, the radar may be switched into a home-on jam mode (if available)—which would be extremely dangerous to the target aircraft.

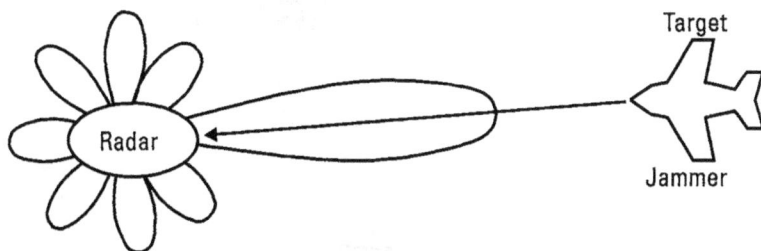

Figure 2.40 Self-protection jamming geometry.

It makes no sense to use a self-protection jammer on a modern stealth aircraft. This has been likened to sneaking into a dark room carrying a 1,000W light bulb.

Expendable decoys and other off-board EA devices are extremely valuable for the protection of aircraft and ships, particularly with the increasing availability of home-on jam modes in weapons. These will be discussed in Section 2.5.

2.4.6 Cover Jamming

The object of cover jamming is to reduce the quality of the signal output from the enemy receiver. Noise modulation is normally used for the jammer because it fills the entire bandwidth of the receiver and reduces the signal to noise ratio (SNR) in the output, causing the same effect as a reduction in the desired signal power. This may prevent an operator from determining that jamming is taking place.

Cover jamming is almost always used for communication jamming, and is typically used by standoff radar jammers. Figure 2.41 shows the effect of cover jamming on a radar plan position indicator (PPI) scope. Although a skilled operator can often read through the noise to track a target return, it may become impossible for the radar processor to automatically track a target. This will have the desirable effect (from the jammer's point of view) of reducing the number of targets that can be engaged by a weapon system. Thus, the defense network can be more easily saturated.

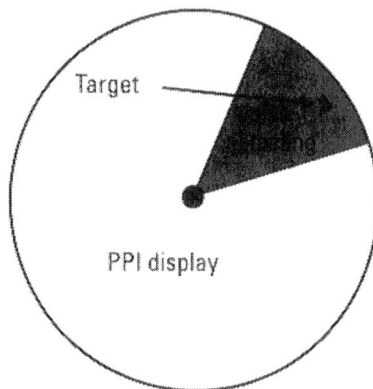

Figure 2.41 Noise jamming on a PPI display.

Barrage Jamming

The simplest type of cover jamming is called *barrage jamming*. It fills a wide frequency spectrum with noise. It covers every frequency at which the communication radio or radar might be operating. The problem with this type of jamming is that most of its power is wasted. Only the noise that falls within the enemy receiver's operating band actually contributes to the J/S. However, the advantage in barrage jamming is that you don't need to know very much about the signal or signals you are jamming.

Spot Jamming

If you know the enemy's operating frequency, you can narrow the jamming noise around that frequency. In the case of a monostatic radar, this information is available if you receive the radar transmissions. One problem, of course, is that it is very hard to receive a communication or radar signal while you are jamming it, since you will be jamming your own (presumably much closer) receiver more effectively than you will be jamming the enemy receiver. Because the enemy may tune the communication radio or radar to avoid jamming (or improve signal quality even though the jamming is not detected), you need to have a look-through scheme.

There are several clever schemes to optimize look-through, but sooner or later you will have to turn off the jammer for a short time while you receive the signal you are jamming. This means that the enemy will receive a few unjammed pulses, or get a short period of unjammed communication. There needs to be a satisfactory compromise between jamming and looking to optimize jammer performance.

Swept Spot Jamming

A compromise noise-jamming technique is to transmit noise over less than the full band that the enemy might be using but to sweep that narrow band across the whole range. This means that the enemy will definitely get some unjammed pulses, but it may prevent the automatic tracking of targets, causing the enemy to use manual tracking by expert operators. This will make the defense network subject to saturation.

2.4.7 Deceptive Jamming

In order to fire on a target, a weapon system must have a radar track on the target to be engaged. The track is a set of range and angle values that allows a prediction of the target's future position. A deceptive jammer causes a radar processor to believe that a target is somewhere else. It thereby breaks the tracking lock of the radar by pulling the radar off in range or angle. Because deceptive jamming requires precise (submicrosecond) knowledge of the parameters of the radar signal arriving at the target, it is normally only applicable to self-protection jamming. There are 10 main types of deceptive jamming:

1. Range gate pull-off (RGPO);
2. Inbound range gate pull-off (RGPI);
3. Cover pulses;
4. Inverse gain;
5. AGC jamming;
6. Formation jamming;
7. Blinking;
8. Cross-polarization;
9. Cross-eye;
10. Terrain bounce.

All of these techniques are described in detail in the EW textbooks listed at the beginning of this chapter. We will discuss just one (range gate pull-off) to give you a feel for how deceptive jamming works.

Range Gate Pull-off

Figure 2.42 shows the signals generated by a range gate pull-off jammer. Figure 2.43 shows what is happening in the radar's processor. A radar, in

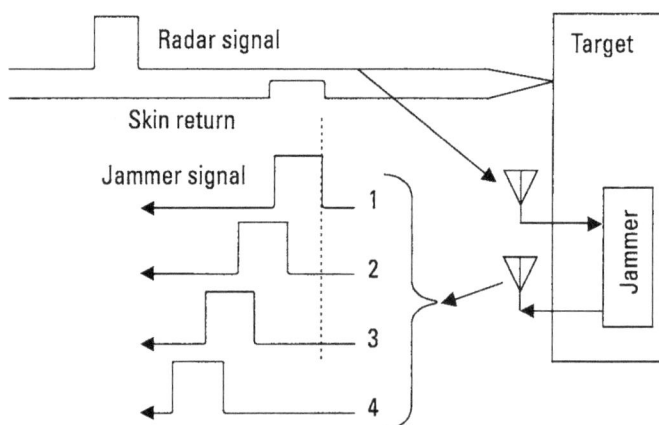

Figure 2.42 Range gate pull-off jammer signals.

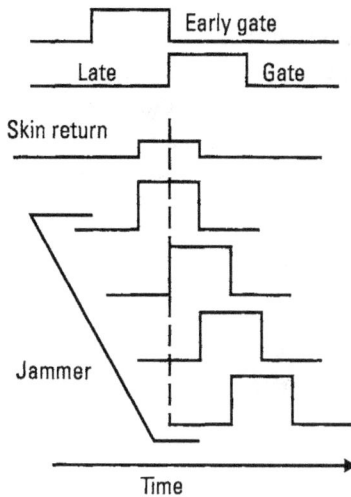

Figure 2.43 Effect of range gate pull-off on radar.

order to maintain range track on a target, has an early gate and a late gate that are centered on the return pulse reflected from the target. The radar determines the range to the target by centering the division between these two gates on the return pulse, equalizing the energy in the two gates. If the target moves farther away, the round-trip time for the pulse will increase.

The deceptive jammer transmits a stronger pulse synchronized with the pulse it receives from the radar. Then the jammer delays its pulse, gradually increasing the amount of delay. The radar sees more energy in its late gate and delays the early/late border out to compensate. Eventually the radar's resolution cell (see Section 2.1.4) moves completely away from the target. This causes the radar to lose lock on the target.

2.4.8 Deceptive Communications Jamming

The equivalent of deceptive jamming in communications is called *spoofing*. This involves the transmission of false information to the enemy in a way that seems to be coming from his own tactical net. This is somewhat dangerous because if the enemy detects the spoofing, he knows what you know about his communication protocol.

Effective spoofing requires that the enemy think the spoofing is the real thing. Any authentication procedures in place must be copied, and if there is encryption, spoofing gets exceptionally hard. One technique is to record enemy transmissions from the commander you want to imitate. Then use a spoofing synthesis system to mix up the words and syllables. The system will also allow you to change the voice tone, add emphasis, imitate excitement, add poor transmission characteristics, and so on. Then, you send the wrong message at some critical time when the receiving operator is likely to forget authentication while the boss (in the boss's recognizable voice) is shouting to take some immediate action.

2.4.9 Directed-Energy Weapons

Directed-energy weapons include high-energy lasers and high-power microwave weapons designed to physically destroy enemy assets. A great deal of research has been done in these fields under the missile defense system programs, and practical weapons will probably be fielded in the near future. However, there are other types of directed-energy weapons that are here now. These weapons are designed to destroy or temporarily disable enemy sensors and communication, or disable almost anything that is computer controlled (vehicle ignition systems, weapons, analysis systems, etc.).

The lasers in this category [called *low-power lasers* (LPLs)] dazzle IR or laser sensors so that they are desensitized to the signals they must receive to guide weapons. With enough power, these lasers can permanently disable the enemy sensors or even disable the platforms. The power required to disable a sensor (precluding effective guidance) is far less than required to physically destroy an aircraft or missile. A laser capable of actually making a hole is called a *high-power laser* (HPL) or a *high-energy laser* (HEL).

Similarly, high-power microwave signals can be beamed at enemy microwave sensors to saturate them so they cannot receive the signals for which they are designed. With more power, sensitive front-end components can be burned out, permanently disabling the sensor.

The principal advantage of these weapons (for either platform or sensor attack) is that their destructive power travels at the speed of light, so they can engage targets in very short time intervals. Laser beams, which must be very tightly focused, can quickly engage many targets without any delay for reloading.

High-power microwave systems have wider beams, so care must be taken not to inadvertently engage friendly assets (thereby committing electronic fratricide).

2.4.10 Antiradiation Missiles

An important aspect of EW is the physical destruction of enemy SAM sites using antiradiation missiles, as shown in Figure 2.44. If you jam a site, it is gone for a little while, but if you blow it away, it stays gone.

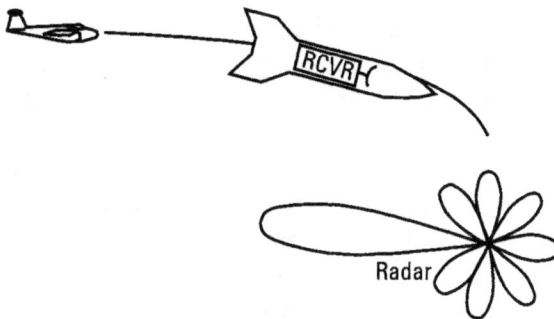

Figure 2.44 Antiradiation missile attack on SAM site.

Destroying the critical part of a SAM site takes precise guidance to a valuable asset that is kept well camouflaged. However, in order to be effective, a radar site must radiate, so homing on its radiation is a dependable way to attack the SAM guidance radar antenna and anything near it. Note that the missile (obviously not to scale in the illustration) has a receiver that is set to the radar frequency and the ability to guide the missile to the location of the radar antenna. Sometimes, pulse time gates are also used for target discrimination. The missile does not "fly down the beam" as commonly portrayed, rather it homes on the radiation, including antenna side lobes.

It is much easier to achieve high accuracy when homing on a target than guiding a weapon to it from a distance. This is because the errors tend to be angular and any given angular error produces less and less miss distance as the range decreases.

The high-speed antiradiation missile (HARM) is carried by several types of aircraft, and can attack SAM sites from beyond their lethal range. This missile system has several modes. It is most effective when it is targeted against a specific site; however it also has an operating mode in which it can be launched without a specific target and attack the highest-priority emitter that comes on the air during its flight. When enemy SAM crews know that there are aircraft in the area capable of firing HARMs, they tend to turn off. Thus, the real value of an antiradiation missile is probably to cause enemy radars to not radiate. Thus, one HARM can, in effect, shut down several SAM sites just by being around.

2.4.11 Infrared Countermeasures

Heat-seeking antiaircraft missiles are among the most deadly weapons faced in modern warfare. They are used in air-to-air combat and as surface-to-air weapons—including inexpensive shoulder-fired missiles. These weapons use passive guidance—homing on the hot parts of jet engines or the jet plume. The IR missile seeker is shown in Figure 2.45. It has a rotating reticule with a pattern of clear and opaque segments that cause a pattern of IR pulses to fall on the IR sensor. The missile analyzes the pulse pattern to determine the steering corrections it must make to home on the target.

The primary countermeasure against heat-seeking antiaircraft missiles has been the launching of flares by the attacked aircraft. Because the flare is much hotter than the heat source the missile is homing on, it captures the missile seeker and leads the attacking missile away from the aircraft, as shown in Figure 2.46. The latest missile trackers have so-called two-color discrimination, which allows them to determine the temperature of the target being

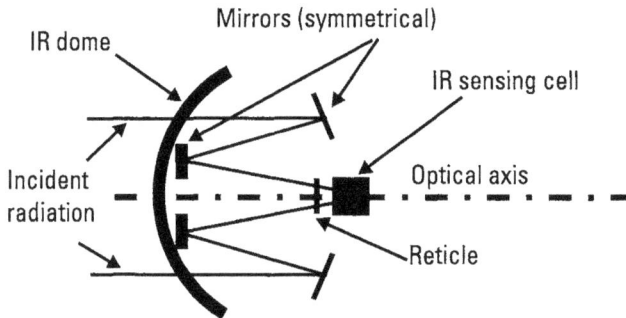

Figure 2.45 IR missile seeker.

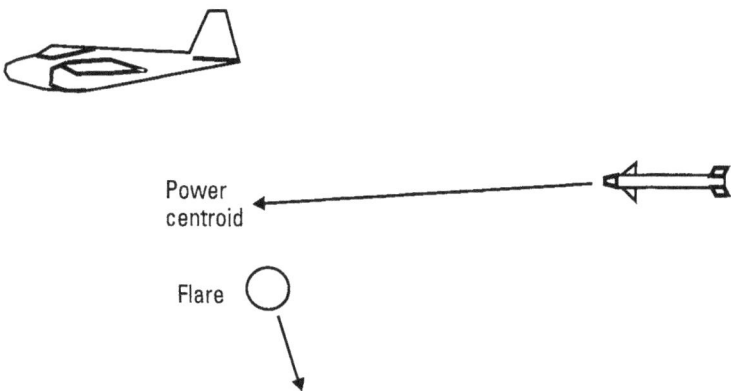

Figure 2.46 Flare drawing missile away from target.

tracked. Since the flares are much hotter than the tracked aircraft, these missiles will discriminate against the flares, requiring much more sophisticated countermeasures.

There are also IR jammers that can be mounted on aircraft and other military vehicles. They generate flashing IR signals in the range of the pulses that reach the IR sensor in the missile from its rotating reticule. These false pulses cause the missile to make improper guidance decisions.

2.4.12 Chaff

Chaff is one of the earliest radar countermeasures used. It consists of strips of aluminum foil or metalized fiberglass with just the right length to optimally reflect radar signals. When a wide range of radar frequencies must be countered, the chaff is cut to a pattern of different lengths to optimally cover the necessary frequency range. Chaff causes radars to receive many returns so that they cannot properly track their targets

Chaff is deployed from pods, from chaff cartridges on aircraft, and from rocket-launching tubes on ships. It can be laid in a corridor through which aircraft fly, as in Figure 2.47, or it can be released in bursts to break the lock of tracking radars, as shown in Figure 2.48.

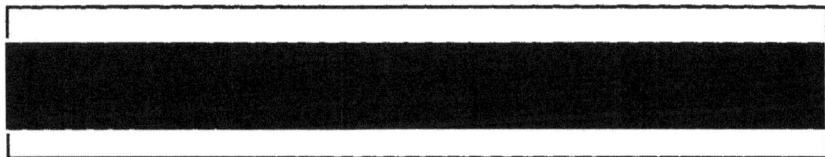

Figure 2.47 Chaff corridor protecting friendly aircraft.

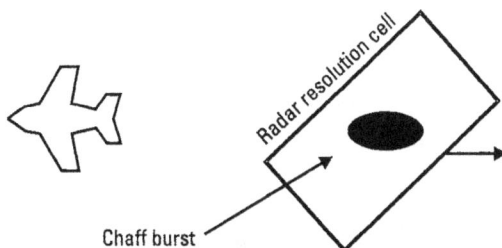

Figure 2.48 Break-lock chaff.

For ship protection, chaff rockets are launched from deck launchers in response to information from the shipboard ESM system. The chaff rockets burst at the optimum elevation above the water to maximize the effective RCS of the resulting chaff cloud.

Antiship missiles are actively guided, using onboard radars that turn on when the missile is approximately 10 km from the ship. The chaff clouds are placed either for distraction or seduction. Distraction chaff is placed in an array some distance from the ship (as shown in Figure 2.49) so that when the missile radar turns on and tries to acquire the ship, it will lock on to a chaff cloud instead. A seduction chaff cloud is fired after the missile radar has acquired the ship. It is placed close to the ship (within the missile radar's resolution cell), as shown in Figure 2.50. The chaff cloud causes the radar to switch its tracking lock to the cloud and is then carried away from the ship by the wind and the ship's motion. The placement is carefully picked (by a computer) to cause the maximum missile miss distance.

2.5 Decoys

Decoys have two classic military applications. The first is to hide important friendly assets among a large number of decoys. The second is to cause an enemy to waste valuable weapons to destroy a low-value decoy. There are now two more important decoy missions: to cause an enemy to expose his defensive assets and to lure a guided weapon away from a protected friendly asset. In any of these applications, it is important that the enemy not be able to easily distinguish the decoy from the real thing.

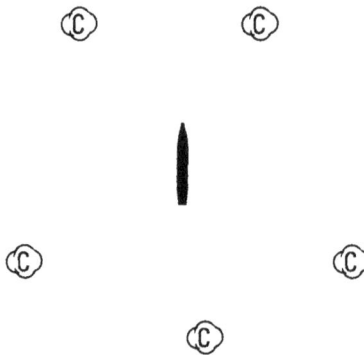

Figure 2.49 Distraction chaff.

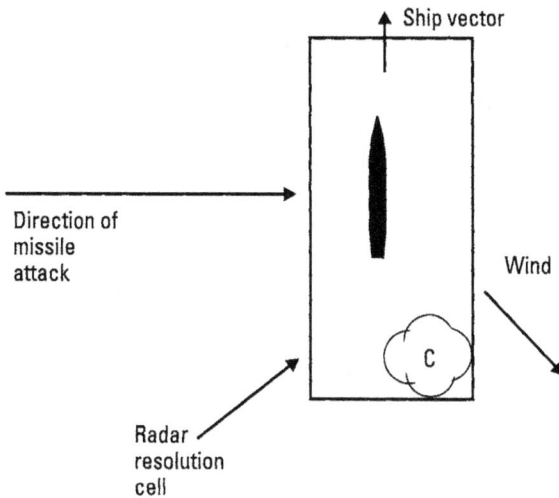

Figure 2.50 Seduction chaff.

Let us focus on applications involving the hostile electronic-tracking of friendly assets. In these cases, the decoy must provide the same return signal that would be produced by the skin return from the protected asset. Since decoys are typically smaller than the assets they protect, they must be made extremely efficient reflectors to produce large RCS in small physical space. This is accomplished passively by use of corner reflectors. A corner reflector returns up to 100 times as much energy (back toward the radar) as a cylinder of the same size.

Active electronic decoys can create enhanced RCS by amplifying received signals and retransmitting them back to the radar, as shown in Figure 2.51. The gain is related to the RCS by the following formula:

$$\sigma = G \lambda^2 / 4\pi$$

where

σ = the RCS;

λ = the radar wavelength;

G = the decoy gain (combined antenna and amplifier gains).

We will use a different form of this same equation in dB form in Chapter 4.

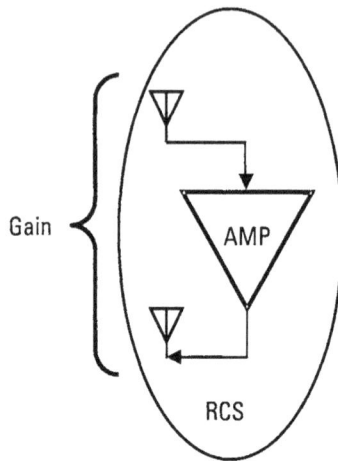

Figure 2.51 Active decoy simulating RCS.

Typical EW-related applications involving each of the four decoy missions are listed below.

- Hide in a crowd:
  - Employed are air-launchable rockets that fly like an aircraft and have enhanced RCS to match the aircraft.
  - Strategic missiles release many decoys that have payloads capable of imitating even the subtle target-size responses analyzed by the most sophisticated radars.
  - Decoys are used for ship defense in the same way described above for distraction chaff.

- Cause enemy to waste valuable weapons:
  - Decoy missile sites, emitting simulated signals, draw million-dollar antiradiation missiles.

- Cause enemy to expose defensive assets:
  - Decoys flying like real aircraft and looking exactly like them to search radars force an enemy to turn on tracking radars. The tracking radars are then located and engaged by antiradiation missiles or smart bombs.

- Seduce radar-guided weapons away from protected assets:
  - Airborne-expendable decoys are dropped behind aircraft to capture the tracking lock of air-to-air missiles to draw them away from the protected aircraft.
  - Maneuverable platforms (usually helicopters or unmanned aerial vehicles) with decoy payloads capture the tracking lock of antiship missiles to draw them away from protected ships.

2.6 Electronic Protection

Electronic protection (EP) is part of the basic systems that enemy electronic countermeasures attempt to make ineffective. This preserves the use of the electromagnetic spectrum for friendly forces. EP was previously called *electronic counter-countermeasures* (ECCMs). This is a very large field, so we will look at only a few typical cases.

Radars have EP features built in to make them less vulnerable to jamming, or particular types of jamming. For example, the Dicke-Fix receiver is a feature added to radars to prevent jamming by the technique called *AGC jamming*. A second radar example is the use of leading-edge tracking to overcome range gate pull-off jamming as discussed in Section 2.4.7.

Communication systems that use any type of low-probability-of-intercept (LPI) technique are doing so in part to make communication jamming much less effective. This is an excellent example of an EP technique.

3

Math for Simulation

There are four math concepts outlined in this chapter (all higher-level math above algebra is explained). The first is dB math—required for the presentation of propagation-theory formulas. The second is spherical trigonometry—to develop EW engagement models, you have to be able to rotate vectors in space using spherical trig. The third is the Poisson equation—in order that you will be able to determine the probabilities required for the design of emulation devices. The fourth is digitization—because almost everything is digital these days.

Note: Those readers fluent in these subjects may feel free to skip this chapter; a perusal of the tables and figures is sufficient for review.

3.1 About dB

In radio propagation calculations, an integral part of EW modeling, we spend a lot of time manipulating widely varying signal strength numbers. Decibel (or dB) units are commonly used in dealing with signal strength values. EW modeling also requires dealing with antenna specifications and RCSs, both of which are usually expressed in dB-based units. Finally, there are a number of easy-to-use formulas relating to radio link and radar performance prediction. The inputs and outputs of these formulas must often be in dB form. Thus, a firm understanding of dB numbers and formulas is important; they will be used extensively through this book, as they are in all aspects of EW modeling and simulation.

Any number expressed in dB is logarithmic, which makes it convenient to compare values that may differ by many orders of magnitude. (Note that numbers in non-dB form are called *linear* in this text to differentiate them from the logarithmic dB numbers.) Numbers in dB form also have the advantage of being easy to manipulate in the following ways:

- To multiply linear numbers, you add their logarithms.

- To divide linear numbers, you subtract their logarithms.

- To raise a linear number to the nth power, you multiply its logarithm by n.

- To take the nth root of a linear number, you divide its logarithm by n.

To take maximum advantage of this convenience, it is common practice to put numbers in dB (i.e., logarithmic) form as early in the process as possible, and to convert them back to linear forms as late as possible (if at all). In many cases, the most commonly used forms of answers remain in dB, so we need not convert back to linear forms.

It is important to understand that any value expressed in dB units must be a *ratio* (which has then been converted into logarithmic form). Common examples include the following:

- Amplifier gain (i.e., the ratio of the output signal strength to the input signal strength);

- Antenna gain (also treated like an amplification ratio, but with some qualifiers);

- Losses (i.e., signal attenuation ratio) when passing through

  - Cables;

  - Switches (the off position, of course, has much more attenuation than the on position, but the on position still has some loss);

  - Power dividers (i.e., ratio of signal power at each output port to input power);

  - Filters;

  - Propagation media (i.e., transmission through space or through atmosphere).

To create useful equations in dB form, it is necessary to express absolute values as dB numbers. Signal strength in units of dBm is the most common example. Since dB values must always be ratios, a trick is required. The trick is to calculate the ratio of the desired absolute value to some fixed value and then convert that ratio to dB form. For example, signal strength in dBm is the dB form of the ratio of that signal strength to 1 mW. (More on this later.)

3.1.1 Conversion to dB Form

The basic formula for conversion into dB is

$$\text{Ratio (in dB)} = 10 \log(\text{linear ratio})$$

For example, 2 (the ratio of 2 to 1) converts to dB form as

$$10 \log(2) = 3 \text{ dB}$$

(Of course, it is actually 3.0103 dB rounded to 3 dB.) Then, ½ (i.e., 0.5) becomes

$$10 \log(0.5) = -3 \text{ dB}$$

To make the conversion the easy way (using a calculator), a scientific calculator with log and 10^x functions must be used.

To convert into dB, do the following:

1. Enter the linear ratio (for example, 2).
2. Press the log key.
3. Multiply by 10.
4. Read the answer in dB (3.0103, which rounds to 3).

To convert back from dB form to linear form, the formula is as follows:

$$\text{Linear Ratio} = 10^{(\text{ratio in dB}/10)}$$

For example:

$$10^{(3/10)} = 10^{(0.3)} = 2$$

Once again, the easy way, using the calculator—to convert dB values back to linear form, do the following:

1. Enter the dB value (for example, 3).
2. Divide by 10 (then hit "=" to get the value onto the display).
3. Press the 10^x key (on most scientific calculators, this requires hitting the second function key and then the log key).
4. Read the answer as the linear ratio (1.99526, which rounds to 2).

The 10^x function is sometimes called the *antilog function* (i.e., second function, log) so you could say that the conversion back to linear form from dB form is by taking the antilog of the dB value divided by 10.

$$\text{Linear ratio} = \text{antilog (dB number/10)}$$

3.1.2 Absolute Values in dB Form

As stated above, the most common example of an absolute value expressed in dB form is signal strength in dBm. This is the ratio of the signal power to 1 mW—converted to dB form exactly as shown in Section 3.1.

Note: dBm is a particularly important unit because many important formulas used in EW modeling either start or end (or both) with dBm values of signal strength.

For example, converting 4W to dBm:

$$4W = 4,000 \text{ mW}$$

$$10 \log(4,000) = 36 \text{ dBm}$$

And of course:

$$10^{36/10} = 10^{3.6} = 4,000 \text{ mW} = 4W$$

or

$$\text{Antilog} (36 / 10) = 4,000 \text{ mW} = 4\,W$$

3.1.3 dB Forms of Equations

dB-form equations use absolute numbers (often in dBm) and ratios (in dB). A typical equation includes only one element in dBm on each side (modified by any number of ratios in dB), or differences of two dBm values (which become dB ratios). One of the simplest dB-form equations is illustrated by the amplifier in Figure 3.1, which multiplies input signals by a gain factor. The linear form of the amplifier equation is

$$P_O = P_I \times G$$

where P_O is the output power, P_I is the input power, and G is the gain of the amplifier.

Both power numbers are in linear units (for example, milliwatts), and G is the gain factor in linear form (for example, 100). If the input power is 1 mW, an amplifier gain of 100 will cause a 100-mW output signal.

By converting the input power to dBm and the gain to dB, the equation becomes

$$P_O = P_I + G$$

The output power is now expressed in dBm. Using the same numbers, 1 mW becomes 0 dBm, the gain becomes 20 dB, and the output power is +20 dBm. (This can be converted back to 100 mW in linear units if required.)

This is a very simple case, in which the marginally simpler calculation does not seem worth the trouble to convert to and from dB forms. But now, consider a typical radio propagation equation. As will be shown in Chapter 4, a transmitted signal is reduced by a spreading loss that is proportional to the square of its frequency (F) and the square of the distance (d) it travels from the transmitting antenna. Thus, the spreading loss is the product of F squared, d squared, and a constant (which includes several terms from the derivation). The formula is

$$L = K \times F^2 \times d^2$$

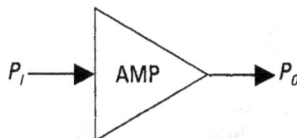

Figure 3.1 Amplifier multiplying input signals by gain factor.

In dB form, F (dB) becomes $10 \log(F)$. F^2 is $2[10 \log(F)]$ or $20 \log(F)$, and d^2 is transformed to $20 \log(d)$ the same way. The constant is also converted to dB form, but first it is modified with conversion factors to allow us to input values in the most convenient units. In this case, K is multiplied by the necessary conversion factors to allow frequency to be input in MHz and distance to be input in kilometers. When the log of this is multiplied by 10, it becomes 32.44, which is commonly rounded to 32. The spreading loss in dB can then be found directly from the expression

$$L_S = 32 + 20 \log(F) + 20 \log(d)$$

where

L_S = spreading loss (in dB);

F = frequency (in MHz);

d = distance (in km).

Most people find this easier to use in practical applications (such as EW modeling). In Chapter 4, dB equations for several one-way and radar-propagation relationships will be presented.

It is important to understand the role of the constant in this type of equation. Since it contains unit conversion factors, this equation only works if you input values in the proper units. In this book, the units for each term are always given right below the dB equation. You will probably memorize some of these equations and use them often; be sure you also remember the applicable units.

3.1.4 Quick Conversions to dB Values

Table 3.1 gives some common dB values with their equivalent linear ratios. For example, multiplying a linear number by a factor of 1.25 is the same as adding 1 dB to the same number in dB form (1 mW × 1.25 is the same as 0 dBm + 1 dB; so 1.25 mW = 1 dBm).

This table is extremely useful because it will allow you to make quick determinations of approximate dB values without touching a calculator. Here is how it works.

First, get from one (1) to the proper order of magnitude. This is easy, because each time you multiply the linear value by 10 you simply add 10 dB to its dB value. Likewise, dividing by 10 subtracts 10 dB from the dB value. Then, use the ratios from Table 3.1 to get close to the desired number. For

Table 3.1

Common dB Values

Ratio	dB Value	Ratio	dB Value
1/10	−10	1.25	+1
¼	−6	2	+3
½	−3	4	+6
1	0	10	+10

example, 400 is $10 \times 10 \times 4$. In dB form, these manipulations are 10 dB + 10 dB + 6 dB. Another way to look at the manipulation is as follows: 400 is 20 dB (which gets you to 100) + 6 dB (to multiply by 4) = 26 dB. And 500 is approximately 30 dB (= 1,000) − 3 dB (to divide by 2) = 27 dB.

Be careful not to be confused by zero dB. A high-ranking government official once embarrassed himself in a large meeting by announcing, "The signal is completely gone when the SNR gets down to zero dB." In fact, a zero dB ratio between two numbers just means that they are equal to each other (i.e., have a ratio of 1).

Table 3.2 shows the power in dBm for various linear power values. This is a useful table, and we will use these values many times in examples in later chapters.

Table 3.2

Signal-Strength Levels in dBm

dBm	Signal Strength	dBm	Signal Strength
+90	1 MW	+20	100 mW
+80	100 kW	+10	10 mW
+70	10 kW	0	1 mW
+60	1 kW	−10	100 μW
+50	100W	−20	10 μW
+40	10W	−30	1 μW
+30	1W	−40	0.1 μW

Other values often expressed in dB form are shown in Table 3.3.

The reader is now given one more chance to be seriously confused about dBs. Voltage ratios are often expressed in dB, but the conversion formula for voltage is

$$\text{Voltage ratio (in dB)} = 20 \log(\text{linear voltage ratio})$$

The basis for this is that the ratio of two power levels is equivalent to the ratio of the squares of two voltage levels (because $P = V^2/R$).

3.2 Spherical Trigonometry

Spherical trigonometry (trig) is an important tool in understanding EW engagements and the operation of some types of EW equipment. In this section, we will run through the basic relationships in spherical triangles and show how to apply spherical trig to several practical EW problems.

3.2.1 The Spherical Triangle

A spherical triangle is defined in terms of a unit sphere; that is, a sphere of radius one (1), as shown in Figure 3.2. The origin (center) of this sphere is placed at the center of the Earth in navigation problems, at the center of the antenna in angle-from-bore sight problems, and at the center of an aircraft or weapon in engagement scenarios. There are, of course, an infinite number of applications, but for each, the center of the sphere is placed where the resulting trigonometric calculations will yield the desired information.

The sides of the spherical triangle must be great circles of the unit sphere; that is, they must be the intersection of the surface of the sphere with

Table 3.3
Common dB Definitions

dBm	= dB value of power/1 mW
dBw	= dB value of power/1W
dBsm	= dB value of area*/1 m²
DBi	= dB value of antenna gain relative to that of an isotropic antenna**

*Commonly used for RCS.
**Defined in Chapter 5.

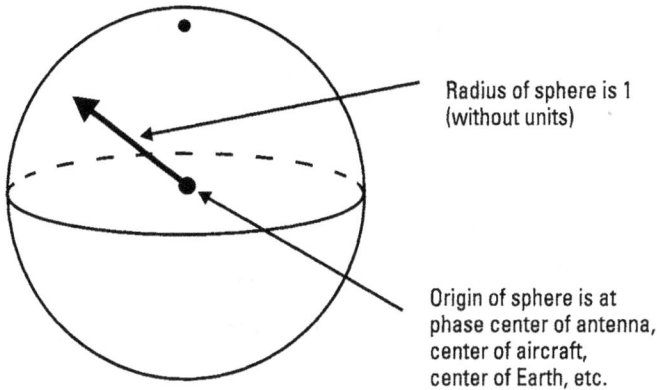

Figure 3.2 Unit sphere.

a plane passing through the origin of the sphere. The angles of the triangle are the angles at which these planes intersect. Both the sides and the angles of the spherical triangle are measured in degrees. The size of a side is the angle the two end points of that side make at the origin of the sphere. In normal terminology, the sides are indicated as lower-case letters, and the angles are indicated with the capital letter corresponding to the side opposite the angle, as shown in Figure 3.3.

It is important to realize that some of the qualities of plane triangles do not apply to spherical triangles. For example, all three of the angles in a spherical triangle could be 90 degrees.

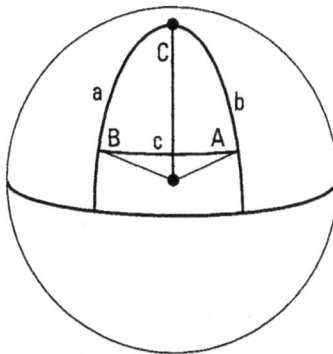

Figure 3.3 Spherical triangle.

3.2.2   Trigonometric Relationships in the Spherical Triangle

While there are many trigonometric formulas, the three most commonly used in EW applications are the law of sines, the law of cosines for sides, and the law of cosines for angles. They are defined as follows:

- Law of Sines for Spherical Triangles

$$\frac{\sin a}{\sin A} = \frac{\sin b}{\sin B} = \frac{\sin c}{\sin C}$$

- Law of Cosines for Sides

$$\cos a = \cos b \cos c + \sin b \sin c \cos A$$

- Law of Cosines for Angles

$$\cos A = - \cos B \cos C + \sin B \sin C \cos a$$

Side a can, of course, be any side of the triangle you are considering, and A will be the angle opposite that side. Those readers who survived high school trig will note that these three formulas are similar to equivalent formulas for plane triangles.

$$\frac{a}{\sin A} = \frac{b}{\sin B} = \frac{c}{\sin C}$$
$$a^2 = b^2 + c^2 - 2bc \cos A$$
$$a = b \cos C + c \cos B$$

3.2.3   The Right Spherical Triangle

As shown in Figure 3.4, a right spherical triangle has one 90-degree angle. This figure shows the way that the latitude and longitude of a point on the Earth's surface would be represented in a navigation problem, and many EW applications can be analyzed using similar right spherical triangles.

Right spherical triangles allow the use of a set of simplified trigonometric equations generated by Napier's rules. Note that the five-segmented disk in Figure 3.5 includes all of the parts of the right spherical triangle except the 90-degree angle. Also note that three of the parts are preceded by the prefix "co-." This means that the trigonometric function of that part of the

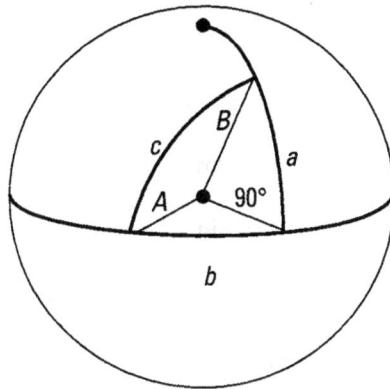

Figure 3.4 Right spherical triangle.

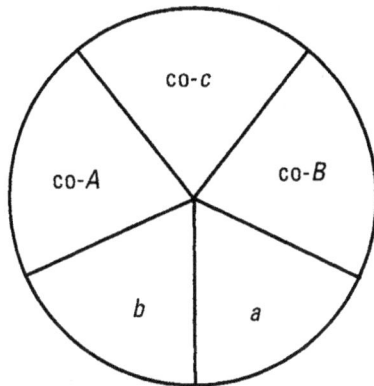

Figure 3.5 Napier's rules chart.

triangle must be changed to the cofunction in Napier's rules (e.g., sine becomes cosine).

Napier's rules are as follows:

- The sine of the middle part equals the product of the tangents of the adjacent parts. (Remember the "co-"s.)
- The sine of the middle part equals the product of the cosines of the opposite parts. (Remember the "co-"s.)

A few example formulas generated by Napier's rules are as follows:

$$\sin a = \tan b \ \cotan B$$

$$\cos A = \cotan c \ \tan b$$

$$\cos c = \cos a \ \cos b$$

$$\sin a = \sin A \ \sin c$$

As you will see when they are applied to practical EW problems, these formulas greatly simplify the math involved with spherical manipulations when you can set up the problem to include a right spherical triangle.

3.2.4   EW Applications of Spherical Trigonometry

Spherical trig is one way to deal with three-dimensional problems, the advantage being that it deals with spatial relationships from the point of view of sensors.

The following examples were chosen because of their relationship to practical EW problems that are highly likely to show up in modeling and simulation applications.

Doppler Shift

Both the transmitter and the receiver are moving (perhaps mounted on two aircraft). Each has a velocity vector with an arbitrary orientation. The Doppler shift is a function of the rate of change of distance between the transmitter and the receiver. What is the rate of change of range between the transmitter and the receiver as a function of the two velocity vectors?

To find the rate of change of distance between the transmitter and the receiver, it is necessary to determine the angle between each velocity vector and the direct line between the transmitter and the receiver. The rate of change of distance is then the transmitter velocity × the cosine of this angle (at the transmitter) + the receiver velocity × the cosine of this angle (at the receiver).

The complexity in the problem comes from the way we define the locations of the transmitter and receiver and the orientation of their velocity vectors.

Let us place the transmitter and receiver in an orthogonal coordinate system in which the y axis is north, the x axis is west, and the z axis is up, as shown in Figure 3.6. The transmitter is located at X_T, Y_T, Z_T and the receiver is located at X_R, Y_R, Z_R. The directions of the velocity vectors are then the elevation angle (above or below the x, y plane) and azimuth (the angle clockwise from North in the x, y plane), as shown in Figure 3.7.

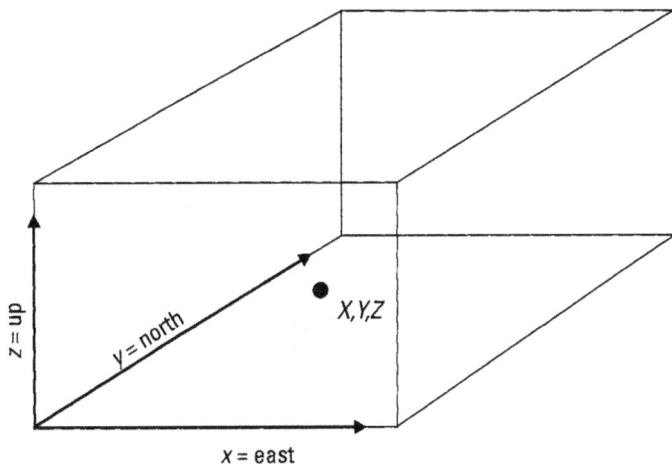

Figure 3.6 Orthogonal coordinate system.

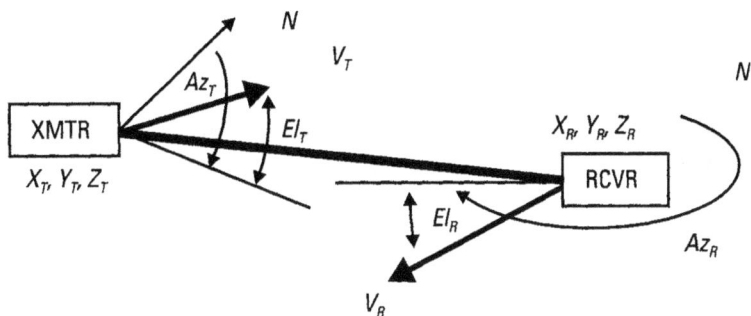

Figure 3.7 Orientation of velocity vectors.

We can find the azimuth and elevation of the receiver (from the transmitter) using plane trigonometry.

$$Az_R = \arctan\left[(X_R - X_T)/(Y_R - Y_T)\right] \qquad (3.1)$$

$$El_R = \arctan\left\{(Z_R - Z_T)/SQRT\left[(X_R - X_T)^2 + (Y_R - Y_T)^2\right]\right\} \qquad (3.2)$$

Now consider the angular conversions at the transmitter, as shown in Figure 3.8. This is a set of spherical triangles on a sphere with its origin at the transmitter. N is the direction to North, V is the direction of the velocity

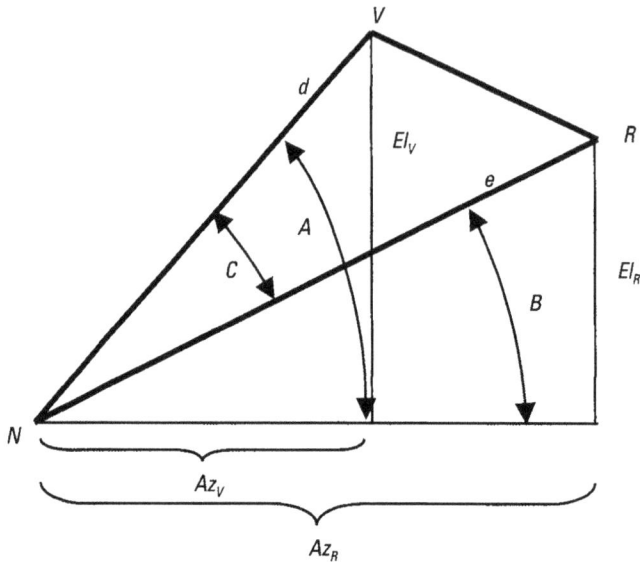

Figure 3.8 Angular conversion of transmitter velocity vector.

vector, and R is the direction toward the receiver. The angle from north to the velocity vector can be determined using the right spherical triangle formed by the velocity vector azimuth and elevation angles. Likewise, the angle from North to the receiver can be determined from the right spherical triangle formed by its azimuth and elevation. From Napier's rules

$$\cos(d) = \cos\left(Az_V\right) \times \cos(El_V) \tag{3.3}$$

$$\cos(e) = \cos\left(Az_R\right) \times \cos(El_R) \tag{3.4}$$

The angles A and B can be determined from (also Napier's rules)

$$\text{ctn}(A) = \sin(Az_V) / \tan(El_V) \tag{3.5}$$

$$\text{ctn}(B) = \sin(Az_R) / \tan(El_R) \tag{3.6}$$

$$C = A - B \tag{3.7}$$

Then, from the spherical triangle between N, V, and R (not a right spherical triangle), using the law of cosines for sides, we find the angle between the transmitter's velocity vector and the receiver:

$$\cos(VR) = \cos(d)\cos(e) + \sin(d)\sin(e)\cos(C) \qquad (3.8)$$

Now the component of the transmitter's velocity vector in the direction of the receiver is found by multiplying the velocity by $\cos(VR)$. This same operation is performed from the receiver to determine the component of the receiver's velocity in the direction of the transmitter.

The two velocity vectors are added to determine the rate of change of distance between the transmitter and receiver (V_{REL}).

The Doppler shift is then found from

$$\Delta f = f V_{REL} / c \qquad (3.9)$$

Elevation-Caused Error in Azimuth-Only DF System

A direction-finding system measures only the azimuth of arrival of signals. However, signals can be located out of the plane in which the DF sensors assume the emitter is located. What is the error in the azimuth reading as a function of the elevation of the emitter above the horizontal plane?

This example assumes a simple amplitude comparison DF. Please note that to apply this approach to a specific type of direction-finding technique, you need to consider the geometry of the way the technique determines direction of arrival (for example the interferometric triangle, as discussed in Section 5.5).

DF systems measure the true angle from the reference direction (typically the center of the antenna baseline) to the direction from which the signal arrives. In an azimuth-only system, this measured angle is reported as the azimuth of arrival (by adding the azimuth of the reference direction to the measured angle).

As shown in Figure 3.9, the measured angle forms a right spherical triangle with the true azimuth and the elevation. The measured angle is labeled M. The true azimuth is determined from Napier's rules in the formula

$$\cos(Az) = \cos(El) / \cos(M) \qquad (3.10)$$

The error in the azimuth calculation as a function of the actual elevation is then

$$\text{Error} = M - \text{arc}\cos\left[\cos(El) / \cos(M)\right] \qquad (3.11)$$

Note that this error peaks at a true 45 degrees of azimuth difference from the reference direction of the DF system.

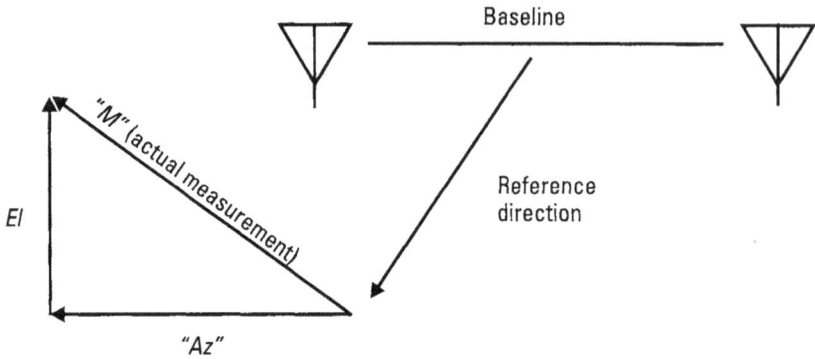

Figure 3.9 Measured angle versus true azimuth and elevation.

Observation Angle in Three-Dimensional Engagement

Given two objects in three-dimensional space, T, for example, is a target and A is a maneuvering aircraft. The pilot of A is facing toward the roll axis of the aircraft, sitting perpendicular to the yaw plane. What are the observed horizontal and vertical angles of T from the pilot A's point of view? This is the problem that must be solved to determine where a threat symbol would be placed on a head-up display (HUD).

Figure 3.10 shows the target and the aircraft in the three-dimensional gaming area of Figure 3.6. The target is at X_T, Y_T, Z_T and the aircraft is at X_A, Y_A, Z_A. The roll axis is defined by its azimuth and elevation relative to the gaming-area coordinate system. The azimuth and elevation of the target from the aircraft location are determined as in (3.1) and (3.2) by

$$Az_T = \arctan\left[\left(X_T - X_A\right) / \left(Y_T - Y_A\right)\right] \qquad (3.12)$$

$$El_T = \arctan\left\{\left(Z_T - Z_A\right) / \mathrm{SQRT}\left[\left(X_T - X_A\right)^2 + \left(Y_T - Y_A\right)^2\right]\right\}$$

$$(3.13)$$

The two right spherical triangles and one nonright spherical triangle of Figure 3.11 allow the calculation of the angular distance from the roll axis to the target (j) using the methodology of (3.3) through (3.7):

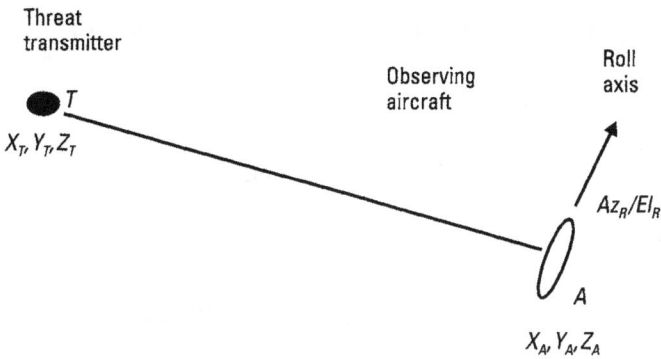

Figure 3.10 Location of target relative to roll axis.

$$\cos(f) = \cos(Az_T) \times \cos(El_T) \qquad (3.14)$$

$$\cos(h) = \cos(Az_R) \times \cos(El_R) \qquad (3.15)$$

$$\operatorname{ctn}(C) = \sin(Az_T) / \tan(El_T) \qquad (3.16)$$

$$\operatorname{ctn}(D) = \sin(Az_R) / \tan(El_R) \qquad (3.17)$$

$$J = 180 \text{ degrees} - C - D \qquad (3.18)$$

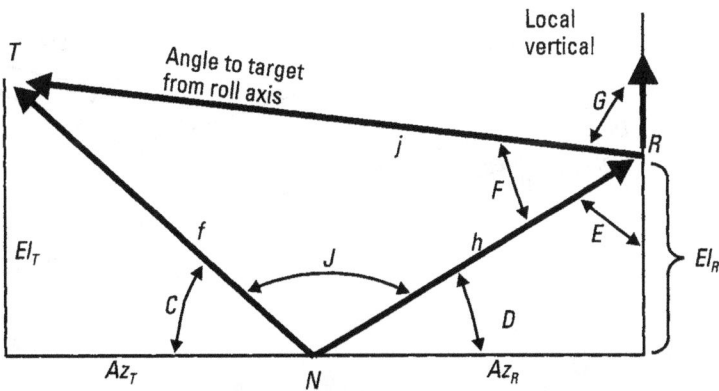

Figure 3.11 Spherical triangles for angle from roll axis to target.

$$\cos(j) = \cos(f)\cos(h) + \sin(f)\sin(h)\cos(J) \qquad (3.19)$$

The angle E is determined from Napier's rules:

$$\text{ctn}(E) = \sin(El_R) / \tan(Az_R) \qquad (3.20)$$

The angle F is determined from the law of sines:

$$\sin(F) = \sin(J)\times \sin(f) / \sin(j) \qquad (3.21)$$

Then the offset angle of the threat from the local vertical at the aircraft is given by

$$G = 180 - E - F \qquad (3.22)$$

Finally, as shown in Figure 3.12, the location of the threat symbol on the head-up display is a distance from the center of the display representing the angular distance j, and an offset from vertical on the head-up display (which, of course, rolls with the aircraft) by the sum of angle G and the roll angle of the aircraft from vertical.

Two important practical points must be understood in order to successfully apply spherical trig to practical problems. First, you need to be careful with trigonometric functions in a spreadsheet (or MathCAD). Their signs change by quadrant, and the tangent function goes infinite at 90 degrees.

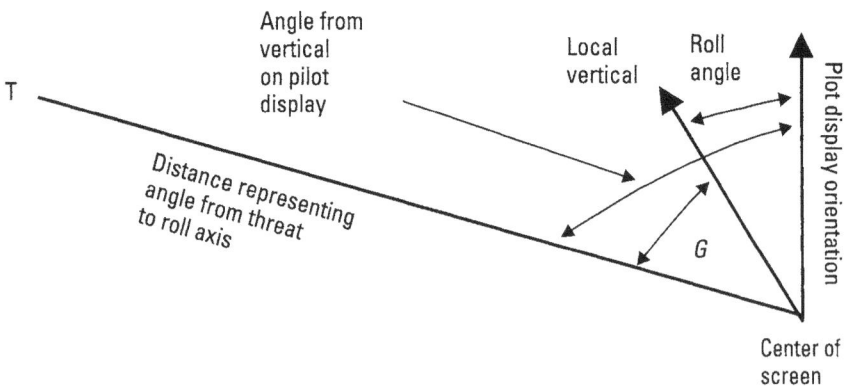

Figure 3.12 Location of threat on HUD.

Second, you can take account of the directions of roll and relative positions of vectors (in the general case) by properly setting up the equations. Neither of these were done in this brief coverage—we just set up the equations for the figures to simplify the discussion of the spherical trig functions.

3.3 The Poisson Theorem

In EW and also in EW simulation, we occasionally find it necessary to predict the number of times that multiple random (or fairly random) events will occur in some time period. The following are important examples:

- The percentage of pulses in a high-density pulse environment that will overlap;
- The number of channels in a dense communications environment that will be found to be occupied when a receiver performs some search operation;
- The probability that a search system will find a randomly tuning signal some number of times during its average message duration.

The Poisson theorem is based on straight probability theory. If there are some percentage of black and the rest white balls in an urn, it predicts the probability that you will pull out exactly k black balls in n tries.

Calculations with this equation on a computer are very straightforward until the number of tries gets large—then the computer memory "maxes out." Therefore, we will cover some ways to get your computer to perform. The equation also works well for determining the probability that at least some number of successes will occur in a fixed number of tries. That is covered as well. It is important to remember that this equation assumes that events are truly random. This can get a little contentious when you are dealing with a large number of cyclic events (like pulse trains). There is a chance that some of these are temporarily harmonically related, but most people let that go if there are large numbers involved.

The basic equation is

$$P_n(k) = \frac{n!}{k!(n-k)!} p^k (1-p)^{n-k}$$

where

$P_n(k)$ = the probability of exactly k successes in n tries;

p = the probability of success in one try.

To fool the computer into doing this calculation, it is often helpful to use the formulas as they simplify for specific numbers of outcomes. This keeps the computer down into numbers it can handle. For example:

$$P_n(0) = (1-p)^n$$

$$P_n(1) = n\,p(1-p)^{n-1}$$

$$P_n(2) = [n(n-1)/2]p^2(1-p)^{n-2}$$

$$P_n(3) = [n(n-1)(n-2)/6]p^3(1-p)^{n-3}$$

An alternate form that is handy in a spreadsheet, using $P_n(3)$ as an example, is:

$$P_n(3) = P_n(2)[(n-2)/3][p/(1-p)]$$

For determining the probability of at least some number of successes in some number of tries, you subtract the "exactly so many" formulas from one. For example, to determine the probability that there will be one or more successes in nine tries, you calculate

$$P_9(\geq = 1) = 1 - P_9(0) = 1 - (1-p)^9$$

If there is a 5% chance of success on each try, there is a 37% chance of at least one success in nine tries. There is a 7% chance of getting at least two successes in nine tries, and .004% chance of at least three.

3.4 Digitization

What follows is a brief look at a very large and important field; just enough to facilitate discussion of the digital aspects of modeling and simulation. For those in need of detailed information about digitization techniques, an excellent resource is *Advanced Techniques for Digital Receivers*, by Phillip E. Pace, (Norwood, MA: Artech House, 2000).

Sampling Rate

It is in general necessary to sample a signal at twice its highest frequency in order to digitize it with adequate quality. However, if there is a frequency band of coverage that is at some higher frequency (for example, an intermediate frequency output) it can be sampled at a rate equal to twice the bandwidth. There are limitations on the absolute frequency that are specified by the device manufacturers.

Another sampling consideration is that if the signal is sampled with two digitizers that have a phase quadrature relationship (to take the I and Q samples of the signal), each can operate at a sample rate equal to the bandwidth.

Number of Quantizing Bits

The level of quantization determines the accuracy with which the digitized signal could later be returned to its original analog form.

Since digitization is binary, there is a fixed number of quantization levels associated with any number of bits in the digital word produced from each sample. For example, 1 bit causes two levels (0 and 1) and 4 bits cause 16 levels (0000, 0001, 0010 . . . on up to 1111). The equation is

$$QL = 2^m$$

where

QL = number of quantizing levels;

m = number of bits per sample.

There is also an equivalent signal-to-noise level (SNR) associated with the number of bits per sample. This is not an actual SNR, but a signal-to-quantization level. It is in fact the ratio of the signal to the distortion of the signal caused by the digitization. The formula is

$$S/N = 3 \times 2^{(2m-1)}$$

where

S/N = the signal-to-quantization (noise) level;

m = is the number of bits per sample.

Table 3.4 lists the number of quantizing levels and the equivalent SNRs for various numbers of bits per sample.

Table 3.4
Quantizing and SNR Versus Bits per Sample

Bits/ Sample	Quantizing Levels	Equivalent SNR (in dB)
1	2	8
2	4	14
3	8	20
4	16	26
5	32	32
6	64	38
7	128	44
8	256	50
9	512	56
10	1,024	62
11	2,048	68
12	4,096	74

4

Radio Propagation

In order to perform EW modeling, it is necessary to understand the basic elements of radio propagation. This chapter covers the so-called one-way link equation, the radar range equation, and several important link equations derived from them.

4.1 One-Way Link Equation

Figure 4.1 is a basic communication link. Its purpose is to take information from the location of the transmitter and deliver it to the location of the receiver. A link is defined as a transmitter, a receiver, two antennas, and everything that happens between the two antennas—or something that acts like that set of elements. Figure 4.2 shows the signal strength as a signal passes through the link. The path through the link is not to scale; it focuses on what happens in each of the elements from the source of the information to its delivery.

The transmitter outputs a radio frequency signal at some signal strength—the transmitted power (P_T). It is defined in some units derived from watts; in EW it is most common to state the transmitted power in dBm.

Next, the signal passes through the transmitting antenna, which has a gain (in dB). In the diagram, this gain is shown as positive, and we will always add the gain to the transmitter power, even though the antenna gain may be positive or negative in dB. It is important to note that the antenna

Figure 4.1 Basic communication link.

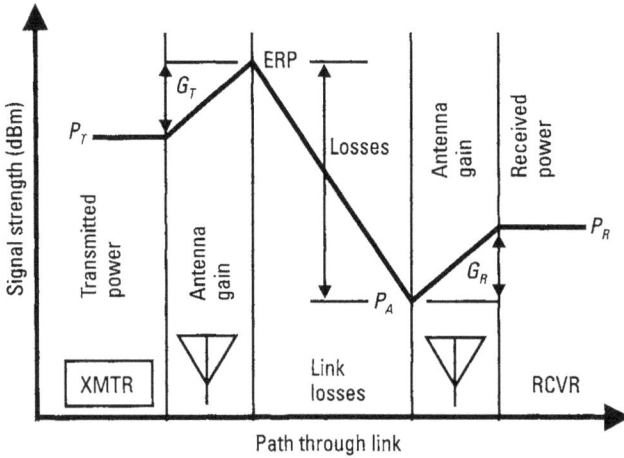

Figure 4.2 Signal-strength levels through link.

gain in this chart is the gain in the direction of the receiver—which may or may not be the peak antenna gain. We will discuss the types of antennas and the trade-offs of their parameters in Section 5.1.

The signal strength at the output of the transmitting antenna (in the direction of the receiver) is called the *effective radiated power* (ERP) and is stated in dBm or dBW. The ERP is the sum (in dB) of the transmitter power and the transmit antenna gain (again, in the direction of the receiver). The formula for ERP is

$$ERP = P_T + G_T$$

where

　　　ERP = effective radiated power (in dBm);

　　　P_T = transmitter power (in dBm);

　　　G_T = transmitting antenna gain in dB;

Next, the signal must propagate through the propagation medium—usually the atmosphere or space—to the receiving antenna. Several types of losses occur in the propagation medium. The primary loss is simply a function of the distance between the two antennas. This is called the *space loss* or *spreading loss*. There are also losses from atmospheric absorption and from rain. If there is not clear line of sight between the transmit and receive antennas, there will also be losses caused by refracting around the curvature of the Earth or over ridge lines. Formulas for each of these functions usually yield transmission loss figures (in dB) between isotropic (or zero dB gain) antennas. Note that we take link losses to be positive numbers of dB, so we can subtract them in equations to reduce the signal strength.

The level of the signal arriving at the receiver location is sometimes called P_A (for power at the receiving antenna). This is a useful term in EW modeling because it allows the isolation of propagation losses (i.e., from ERP to P_A) from the rest of the link effects.

At the receiving location, the signal is accepted by the receiving antenna, which has a gain in dB (also either positive or negative). The receiving antenna gain in this chart is the gain in the direction of the transmitter—not necessarily the antenna's peak gain.

The power leaving the receiving antenna is called the *received power* (P_R). It is the input to the receiving system, and is stated in dBm.

This chart, along with many of the propagation manipulations we do in EW, assumes that the signal strength can be stated in dBm through the whole link. This is convenient, but is not literally true. Out in the transmission medium between the two antennas, there is only field strength, often stated in microvolts per meter. In order to stay in dBm, we use the artifice of an ideal 0-dB-gain antenna. The output of this ideal antenna would be in dBm, and that is the assumed level we use at any point in the propagation path (including ERP and P_A).

The one-way link equation can be stated as

> Transmitter power
> + transmitting antenna gain
> − link losses
> + receiving antenna gain
> = received power

The chart in Figure 4.2 is, in effect, the link equation.

An example follows:

$$\text{Transmitter power} = 10W(= 40dBm)$$
$$\text{Transmit antenna gain} = 6dB$$
$$\text{Link losses} = 100dB$$
$$\underline{\text{Receiving antenna gain} = -2dB}$$
$$\text{Received power} = -56dBm$$

Note that a detailed analysis of a specific link sometimes has separate entries for internal losses (such as cable losses to antennas), which we assume here to be included in the other component values.

4.2 Propagation Losses

This section deals with spreading loss, atmospheric loss, and rain loss because non-line-of-sight losses are seldom dealt with in modeling, we will not discuss them. Typically, any line-of-sight path that drops below the terrain or beyond the four-thirds Earth radio horizon is assigned an infinite loss in EW modeling and simulation. Actual line-of-sight losses are quite probabilistic and situation specific; they are normally dealt with only in detailed propagation simulations. The communications handbook available from IEEE (ISBN: 0-8493-8349-8) covers this to an adequate level for most people needing to include it in a specific situation model.

4.2.1 Spreading Loss

The most important loss factor, because it accounts for most of the link loss, is the spreading loss L_S. The following equation is for the spreading loss between 0 dB gain (also called *isotropic*) antennas. It is a function of distance squared and frequency squared.

The most commonly used form of this equation is

$$L_S = 32 + 20 \log(F) + 20 \log(d)$$

where

L_S = spreading loss in dB
F = frequency in MHz;
d = distance in km.

The factor 32 is a constant that is only valid for the stated units (MHz for frequency and km for distance). Like most other dB formulas, any other units used will yield an incorrect answer. Also, the constant is rounded to

1 dB (a common practice). Finally, note that the log used in the formula must be log to the base 10 (which is the log key on a calculator). The constants for common units of link distance are shown in Table 4.1.

Figure 4.3 is a nomograph with which you can quickly determine the spreading loss as a function of frequency and distance. Draw a line between

Table 4.1
Spreading Loss Constants

Distance Units	Equation Constant	Commonly Rounded to
Kilometers	32.44	32
Statute Miles	36.52	37
Nautical Miles	37.73	38

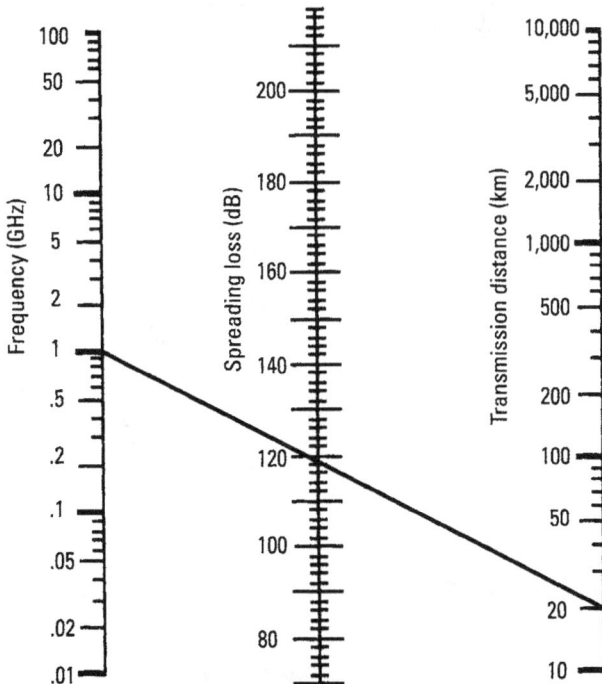

Figure 4.3 Spreading-loss nomograph.

the desired frequency and the desired distance and read the loss in dB from the center scale. The example on the scale shows that 1 GHz at 20 km yields approximately 118.5 dB of spreading loss.

4.2.2 Atmospheric Loss

In the Earth's atmosphere, radio signals are principally absorbed by oxygen and water vapor. Figure 4.4 is a graph of atmospheric absorption versus frequency per kilometer at sea level. To use this graph, draw a line up from the frequency of propagation to the curve, then left to the dB per kilometer. Then simply multiply that value by the number of kilometers between the transmit and receive antennas. Note that this loss is in addition to the spreading loss.

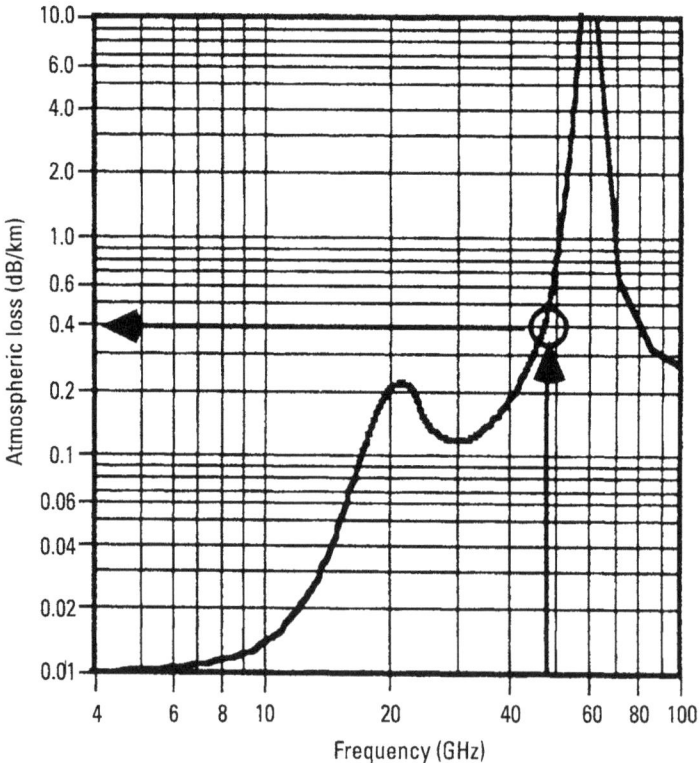

Figure 4.4 Atmospheric loss.

You will notice that there is a peak loss at about 22 GHz. This is from water vapor. The higher peak at 60 GHz is from oxygen. The actual top of this peak is about 20 dB/km. That is why nobody in his or her right mind designs a surface-to-space link at 60 GHz, but there are lots of satellite-to-satellite links near 60 GHz.

4.2.3   Rain Loss

You will have noticed from your own experience that rain is not consistent. In most climates, it will usually rain at a fairly light level for an extended period of time, but may switch to very hard rain for short periods. Of course, in a tropical rain forest, or high in a mountain range, it might rain very hard for hours. Thus, it is only possible to design a communication link to operate with some model of rainfall. Once the model is established, you can trade off various configurations (including frequency selection) to determine the rain performance versus the other qualities the link must have. A typical rainfall model might be like that in Figure 4.5. This is for a 50-km link with light rain over the whole distance, but heavy rain cells totaling 10 km of the link distance (i.e., 40 km of light rain plus 10 km of heavy rain).

The chart in Figure 4.6 is attenuation per kilometer for various defined levels of rainfall versus frequency. Use this chart just like the atmospheric loss chart. At the link frequency, go up to the line, then left to determine the dB per km. Then multiply this value by the number of kilometers of path subject to that level of rain. For the example shown on the chart, the 10 km of heavy rain produces 7.5 dB of loss and the 40 km of light rain produces 1.4 dB for a total of 8.9 dB at 15 GHz.

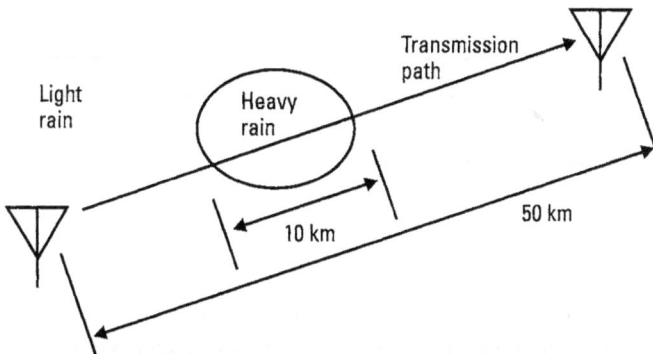

Figure 4.5 Typical rainfall model.

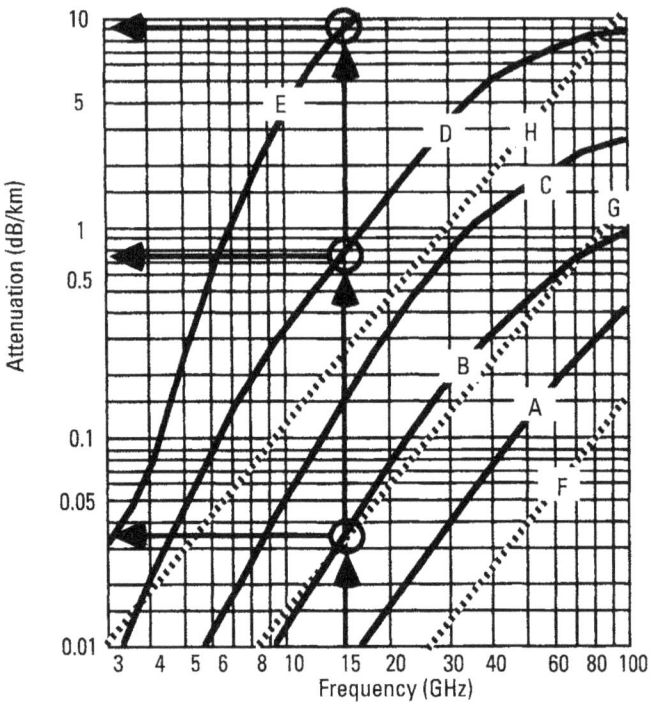

	A	0.25 mm/hr	.01 in/hr	Drizzle
	B	1.0 mm/hr	.04 in/hr	Light rain
Rain	C	4.0 mm/hr	.16 in/hr	Moderate rain
	D	16 mm/hr	.64 in/hr	Heavy rain
	E	100 mm/hr	4.0 in/hr	Very heavy rain

	F	0.032 gm/m^3	Visibility greater than 600m
Fog	G	0.32 gm/m^3	Visibility about 120m
	H	2.3 gm/m^3	Visibility about 30m

Figure 4.6 Attenuation from rain and fog.

4.3   Receiver Sensitivity

The sensitivity of a receiver is the minimum signal it can receive and still do its job. We are concerned here with the sensitivity of what we will call the receiver system. That is one or more receivers and the necessary cabling, switching, and so on so they can accept the output of an antenna.

In Section 4.2, we developed the formulas to predict the received signal strength—that is, the power level of the signal arriving at the receiver system input. If the receiver system sensitivity is equal to or less than that signal strength, the link does its job. If the received signal is greater than the sensitivity level of the receiver, there is said to be *link margin,* which is the numerical difference between the two values (usually expressed in dB).

There are three components to sensitivity, as shown in Figure 4.7. They are kTB, noise figure, and SNR.

kTB

kTB defines the noise in an ideal receiver, where k is Boltzman's constant, T is the temperature in degrees Kelvin (°K), and B is the effective bandwidth of the receiver. Within the Earth's atmosphere, we can assume the temperature to be 290°K, which allows us to use the following formula for virtually all EW applications:

$$kTB = -114 \text{ dBm} + 10 \log(BW)$$

where kTB is in dBm and BW is the effective receiver bandwidth in MHz.

Figure 4.8 is a graph of kTB versus receiver effective bandwidth, in case you would rather use a chart than a formula.

Noise Figure

The system noise figure is the amount of noise that the system adds above kTB, but it is referenced back to the receiver system input (i.e., the output of the antenna). That is sometimes defined as the amount of noise that would

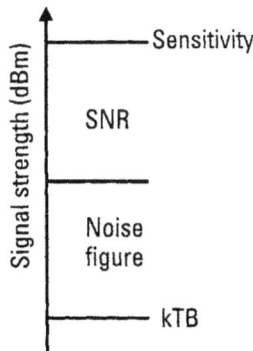

Figure 4.7 Components of sensitivity.

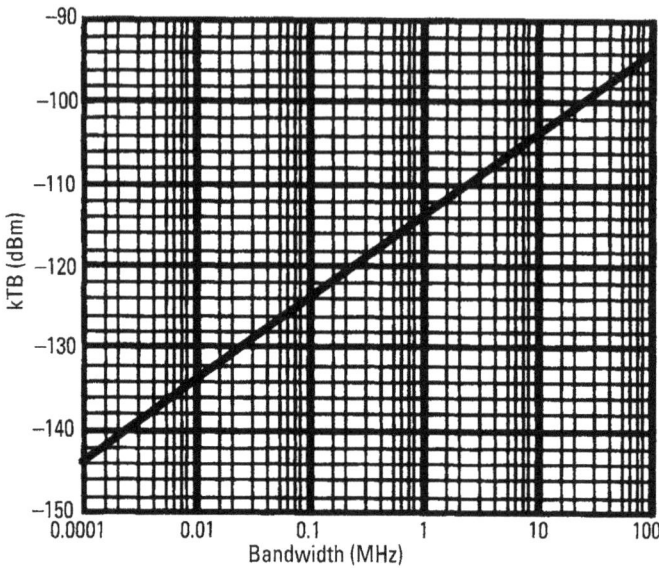

Figure 4.8 kTB versus bandwidth.

have to be added with a noise generator connected to the input to create the measured output noise—if the system were noiseless.

The lower the system noise figure, the better the sensitivity. Factors that reduce system noise figure are the use of a preamplifier, minimizing cable losses between the antenna and the receiver or preamplifier, and using amplifiers and receivers with reduced noise figure. Typical system noise figures vary from 4 to 6 dB if there is a preamplifier close to the antenna to as much as 20 dB if no preamplifier is used.

Signal-to-Noise Ratio

The SNR is the way we quantify the quality of a signal. The required SNR is a function of the type of signal, the information the receiver will recover from the signal, and the output quality required. It should be noted that the SNR we are concerned with is the SNR at the input of the receiver. This is sometimes the SNR of the system audio or video output, but with some modulations it is different. This input SNR is called the *carrier-to-noise ratio* (CNR) or, more correctly the RF SNR.

Table 4.2 shows some typical values for the RF SNR. These are commonly accepted average values, but may vary widely depending on the type of equipment used, the circumstances, and the nature of the information being carried or the analysis being performed by a hostile receiver.

Table 4.2
Required RF SNR for Various Types of Signals

Type of Signal	RF SNR Required (dB)
Broadcast TV picture	40
Voice push to talk AM radio	15
Voice push to talk FM radio	12
Pulse video—Expert operator	8
Pulse video—Automatic computer processing	15
SEI analysis	30 to 50
Radar	13

Sensitivity Calculation Example

If the effective bandwidth of a receiver is 10 MHz, its noise figure is 18 dB, and the signal is a pulse train that is to be automatically computer processed, the receiver sensitivity is

$$(kTB) - 114 \ \text{dBm} + 10 \log(10) = -104 \ \text{dBm}$$
$$+ (\text{Noise figure}) \ 18 \ \text{dB}$$
$$\underline{+ (\text{Required SNR}) \ 15 \ \text{dB}}$$
$$= -71 \ \text{dBm}$$

4.4 Effective Range

The effective range of a link is the range at which the received power equals the receiver system sensitivity. If we consider only spreading loss, the received power is given by the formula

$$P_R = \text{ERP} - 32 - 20\log(F) - 20\log(d) + G_R$$

where

P_R = received power (dBm);
ERP = the transmitter effective radiated power in dBm;

F = frequency (MHz);

d = distance (km);

G_R = receiving antenna gain (dB).

Setting P_R equal to the sensitivity (sens) yields

$$\text{Sens} = \text{ERP} - 32 - 20\log(F) - 20\log(d) + G_R$$

Solving for $20\log(d)$ gives

$$20\log(d) = \text{ERP} - \text{sens} - 32 - 20\log(F) + G_R$$

Remember that sensitivity and ERP are in dBm, frequency is in MHz, and distance is in kilometers.

Solving for d gives

$$d = \text{antilog}\{20\log(d)/20\}$$

You will recall that the antilog is 10 raised to the power within the brackets.

Effective Range Example

If a radar signal has an ERP of 1MW at 10 GHz and the receiver has −71 dBm sensitivity, the receiving antenna has 2dB gain, and the effective range (i.e., the range at which the receiver can adequately receive the radar signal) is calculated as

$$\text{ERP} = 1\text{MW} = +90 \text{ dBm}$$
$$20\log(F) = 20\log(10{,}000) \quad \{1\,\text{GHz} = 1{,}000\,\text{MHz}\}$$
$$= 80$$

$$20\log(d) = 90 \text{ dBm} - (-71\,\text{dBm}) - 32 - 80 + 2 = 51$$
$$d = \text{antilog}\,(51/20) = \text{antilog}\,(2.55) = 355\,\text{km}$$

This particular value (since we chose a radar as the signal) is also called the radar's *detectability range*.

To determine the effective range including atmospheric and rain losses, it is common practice to first estimate the range, then calculate the atmospheric and rain losses (which are functions of range. Then subtract the value of those losses from the value of $20 \log(d)$ and recalculate d. This may have to be repeated a few times to calculate d to adequate accuracy. This is fairly easy to do in a computer.

4.5 Radar Range Equation

The radar range equation is an expression for the received energy in the radar receiver, assuming that that the transmitter and receiver are colocated. It is commonly stated without considering any link losses except spreading loss. The equation is often given in the form

$$E_R = \left[P_T G^2 \sigma \lambda^2 T \right] / \left[(4\pi)^3 R^4 \right]$$

where

E_R = the received energy (W/s)

P_T = the radar transmitter power (W);

G = the radar antenna gain (not in dB);

σ = the redar cross section of the target (m^2);

λ = the wavelength of the radar signal (m);

T = the time that the target is illuminated (sec);

R = the range to the target (m).

This shows only what the radar receiver receives, and is useful in determining radar performance. However, from an EW point of view—particularly an EW modeling and simulation point of view—we need to have a good feeling for what is going on during the entire round trip to the target and back, and we are only interested in the received power. Therefore, we will consider a different form of the radar range equation (which also considers only the spreading loss). This form is in dB units, and is separated into segments as shown in Figure 4.9.

The signal strength at the output of the transmit antenna (P_1) is the ERP; in dB form it is

$$P_1 = P_T + G$$

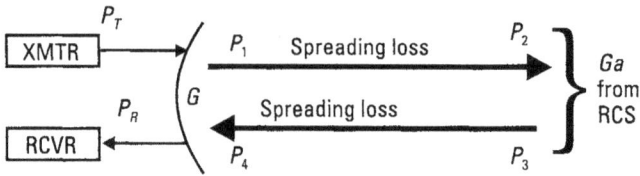

Figure 4.9 Segments of radar range equation.

where

$P_T =$ the transmitter power (dBm);
$G =$ the antenna gain (dB).

The signal strength arriving at the target (P_2) is reduced by the spreading loss $[L_s = 32 + 20 \log (F) + 20 \log(R)]$, which makes the signal strength

$$P_2 = P_T + G - 32 - 20 \log(F) - 20 \log(R)$$

where

$P_T =$ the transmitter power (dBm);
$G =$ the antenna gain (dB);
$F =$ the radar's transmit frequency (MHz);
$R =$ the range to the target (km).

At the target, the radar signal is reflected with an efficiency that is dependent on both the RCS and the frequency. The formula for the signal gain (i.e., $P_3 - P_2$) during the target reflection (in dB form) is

$$Ga = -39 + 20 \log(F) + 10 \log(\sigma)$$

where

$Ga =$ the effective reflection gain (dB);
$F =$ the frequency (MHz);
$\sigma =$ the radar side lobe gain (dB).

The radar signal strength (P_3) leaving the target is then defined by the formula

$$P_3 = P_T + G - 71 - 20 \log(R) + 10 \log(\sigma)$$

where

P_T= the transmitter power (dBm);

G= the antenna gain (dB);

R= the range to the target (km);

σ = the RCS (m^2).

You will note that the frequency term is now gone from the equation, because the + 20 log(F) from the RCS gain term offsets the –20 log(F) from the one-way attenuation from the radar to the target.

As the reflected signal returns to the radar location, it is further attenuated by a one-way spreading loss, to make the formula

$$P_4 = P_T + G - 103 - 40 \log(R) - 20 \log(F) + 10$$

Now consider the received signal strength—that is, the power input to the receiver. It is increased by the receiving antenna gain, which is the same as the transmitting antenna gain. So the radar range equation in this dB form is now

$$P_R = P_T + 2G - 103 - 40 \log(R) - 20 \log(F) + 10 \log(\sigma)$$

where

P_R = the signal strength to the radar receiver (dBm);

P_T= the radar transmitter power (dBm);

G = the antenna gain (dB);

R = the range to the target (km);

σ = the RCS (m^2).

This equation is the same as the non-dB form given above—with the time term removed, the wavelength converted to frequency and the unit conversion factors grouped together and placed in dB form.

4.6   Range Limitation from Modulation

The pulse width and pulse repetition interval also limit the range of a radar. The pulse width determines the minimum range. Basically, the reception of the reflected pulse cannot start until the transmission of the transmitted pulse has ended (see Figure 4.10). Therefore, the round-trip distance cannot be less than the distance that would be traveled at the speed of light during the pulse width.

The pulse repetition interval limits the maximum unambiguous range of a radar, as shown in Figure 4.11. If the reflection of one pulse is not received before a second pulse is transmitted, the determination of target range is ambiguous (i.e., which return pulse represents the true range?). Thus, the maximum unambiguous range is the round-trip distance that a signal travels at the speed of light during a pulse repetition interval.

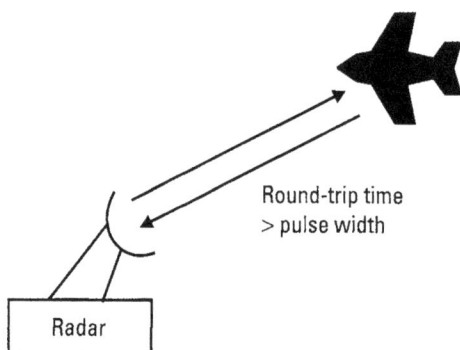

Figure 4.10 Pulse-width limitation of minimum range.

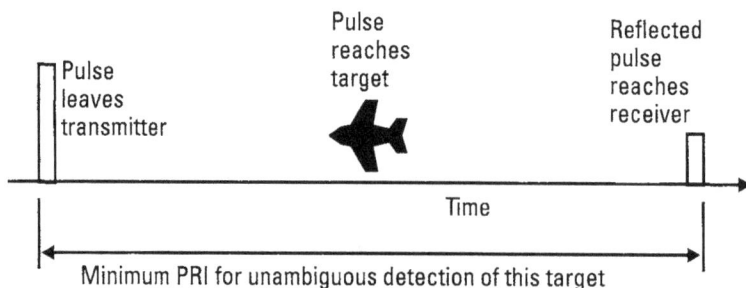

Figure 4.11 PRI limitation of maximum range.

4.7 Radar Detection Range

The radar receiver has a sensitivity that is determined as indicated in Section 4.3. This sensitivity can be plugged into the radar range equation in place of the received power. Then the equation can be solved for the range, to determine the range at which the received power equals the sensitivity of the radar receiver (i.e., the detection range). The equation is as follows:

$$\text{Sens} = P_T + 2G - 103 - 40\log(R) - 20\log(F) + 10\log(\sigma)$$

Then we can solve for $40\log(R)$ as follows:

$$40\log(R) = P_T + 2G - 103 - 20\log(F) + 10\log(\sigma) - \text{sens}$$

We can solve for R by taking

$$R = \text{antilog}\{40\log(R)/40\}$$

For example, if the radar transmit power is 1,000W, its antenna gain is 30 dB, the frequency is 10 GHz, the RCS of the target is 10 m^2, and the sensitivity is –71 dBm:

$$1{,}000\text{W} = +60\text{ dBm}$$

$$40\log(R) = +60\text{dBm} + 2(30\text{ dB}) - 103 - 20\log(10{,}000)$$
$$+10\log(10) - (-71\text{dBm})$$
$$= +60 + 60 - 103 - 80 + 10 + 71 = 18$$

$$R = \text{antilog}\{18/40\} = 2.8\text{ km}$$

This is the detection range of the radar for the conditions in this example.

4.8 Jamming-to-Signal Ratio

The J/S ratio is the ratio of two signal strengths as they arrive at the receiver. Each signal is affected by the same radio propagation parameters we have been discussing. We will present three cases: communication jamming, standoff jamming of a radar, and self-protection jamming of a radar.

4.8.1   Communication Jamming

Communication jamming applies to tactical communication, data links, and any other case in which there is transmission of data from one point to a second remote point. The propagation involved in communication jamming is shown in Figure 4.12.

It is important to remember that the jammer always jams the receiver (as opposed to the transmitter). In tactical communications, where each communicator has a transceiver, it is the link transmitting to the location being jammed that is affected by the jamming. Although tactical communication almost always has 360-degree antennas, we will make provision in the following formulas for different receiving antenna gains to the jammer and the desired signal. The J/S ratio is the received signal strength for the jammer (J) divided by the received signal strength of the desired signal (S).

The received signal level for the jammer is

$$J = \mathrm{ERP}_J - 32 - 20\log(F) - 20\log\left(d_J\right) + G_{RJ}$$

where

J = the jammer signal strength to the receiver (dBm);

ERP_J = the jammer effective radiated power (dBm);

F = the communication frequency (MHz);

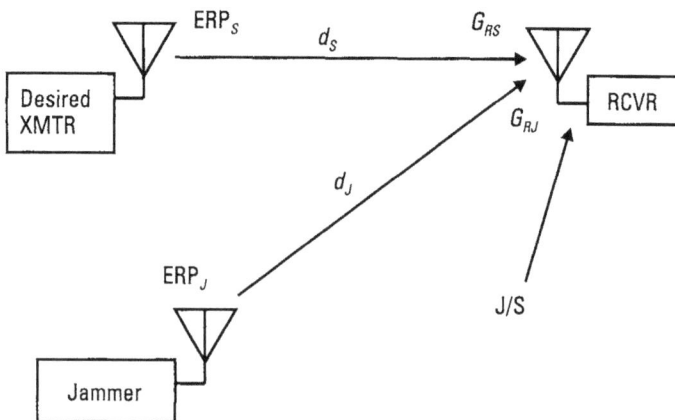

Figure 4.12 Communication-jamming geometry.

G_{RJ} = the receiving-antenna gain in the direction of the jammer (dB);

d_J = the range from the jammer to the receiver (km).

The received signal level for the desired signal is

$$S = \text{ERP}_S - 32 - 20\log(F) - 20\log(d_S) + G_{RS}$$

where

S = the desired signal strength to the receiver (dBm);

ERP = the desired transmitter effective radiated power (dBm);

F = the communication frequency (MHz);

G_{RS} = the receiving antenna gain in the direction to the desired signal transmitter (dB);

d_S = the range from the desired signal transmitter to the receiver (km).

The jamming to signal ratio (J/S) is the difference between these two numbers (J – S)in dB:

$$\text{J/S} = \text{ERP}_J - \text{ERP}_S - 20\log(d_J) + 20\log(d_S) + G_{RJ} - G_{RS}$$

You will notice that the frequency term goes out because both the desired transmitter and the jammer have the same frequency.

For example, if the desired signal ERP is 10W, at 10 km from the receiver, the jammer ERP is 1 kW at 40 km, and the receiving antenna is nondirectional, the J/S would be

$$10\text{W} = +40 \text{ dBm}$$
$$1 \text{ kW} = +60 \text{ dBm}$$
$$20\log(10) = 20$$
$$20\log(40) = 32$$

The two antenna gain terms are equal and thus cancel.

$$\text{J/S} = +60 - 40 - 32 + 20 = 8 \text{ dB}$$

4.8.2 Standoff Jamming of a Radar

Standoff jamming uses the geometry shown in Figure 4.13. The jammer is outside the lethal range of the weapon controlled by the radar, so it is

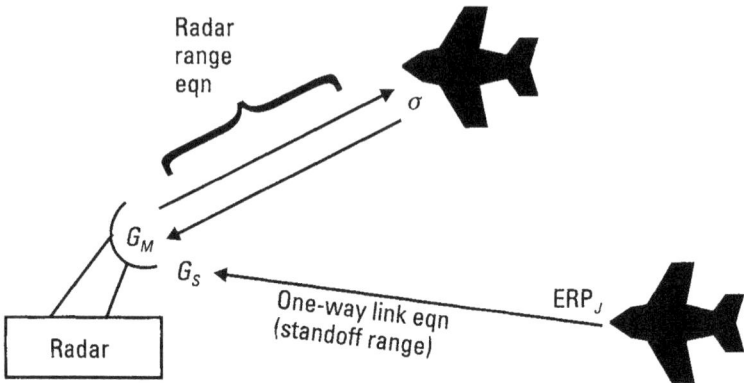

Figure 4.13 Standoff-jamming geometry.

assumed to be at a different range than the radar-to-target distance. The radar antenna has its main beam pointed at the target, so the standoff jammer is assumed to be outside the main beam. We assume that the jammer is received in a side lobe, as explained in Chapter 5. The received power from the skin return of the target is given by the radar range equation. We have changed the antenna gain designation to specify main beam gain:

$$S = P_T + 2G_M - 103 - 40\log(R_T) - 20\log(F) + 10\log(\sigma)$$

where

S = the signal strength of the skin return (dBm);
P_T = the transmitter power (dBm);
G_M = the radar main beam gain (dB);
R_T = the range to the target (km);
F = frequency (MHz);
σ = the RCS (m^2).

The received power from the jammer is given by the formula

$$J = \mathrm{EPR}_J - 32 - 20\log(F) - 20\log(R_J) + G_S$$

where

J = the jammer signal strength to the radar (dBm);

ERP_J = the jammer effective radiated power (dBm);

F = the radar frequency (MHz);

G_S = the radar side lobe gain (dB);

R_J = the range from the jammer to the radar (km).

The J/S ratio is the difference between these two equations (in dB units):

$$J/S = ERP_J - P_T + 71 - 20 \log(R_J) + 40 \log(R_T) - 10 \log(\sigma) + G_S - 2G_M$$

For example, if the radar has a 1,000W transmitter into a 30-dB antenna with 0-dB side lobes, a 10 m² target is at 10 km, and the jammer is at 30 km with a 10,000W ERP:

$$1,000\,W = +60\,dBm, \quad 10,000\,W = +70\,dBm$$
$$40 \log(10) = 40, \quad 20 \log(30) = 30$$
$$J/S = +70 - 60 + 71 - 30 + 40 - 10 + 0 - 60 = 21\,dB$$

As the range to the target decreases, the J/S reduces by the 4th power of the reducing distance. The range at which the radar can establish track in the presence of jamming is called the *burn-through range*.

4.8.3 Self-protection Jamming

When the jammer is located on the target (Figure 4.14), self-protection jamming occurs. This type of jamming can usually achieve much higher J/S ratios because the jammer has the advantage of the radar's main beam antenna gain.

The expression for the signal strength of the skin return is the same as that for standoff jamming, but the formula for the signal strength of the jammer at the radar receiver is somewhat different:

$$J = ERP_J - 32 - 20 \log(F) - 20 \log(R_T) + G_M$$

where

J = the jammer signal strength to the radar (dBm);

ERP_J = the jammer effective radiated power (dBm);

F = the radar frequency (MHz);

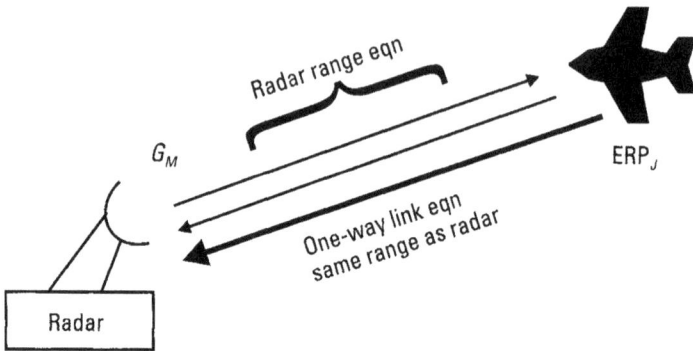

Figure 4.14 Self-protection jamming geometry.

G_M = the radar main beam gain (dB);

R_T = the range from the jammer and target to the radar (km).

The J/S formula can be written in a much simpler form, since the range and the radar antenna gain are the same in both the J and S formulas. J/S is now

$$J/S = ERP_J - P_T + 71 + 20 \log(R_T) - 10 \log(\sigma) - G_M$$

For example, if the radar has a 1,000W transmitter into a 30-dB antenna and a 10 m^2 target is at 10 km with a self-protection jammer with 100W ERP, then

$$ERP_J = 100W = +50 \text{ dBm};$$
$$P_T = 1,000W = +60 \text{ dBm};$$
$$20 \log (R_T) = 20\log (10) = 20;$$
$$10 \log (\sigma) = 10 \log (10) = 10;$$
$$G_M = 30 \text{ dB}.$$

$$J/S = +50-60+71+20-10-30 = 41 \text{ dB}$$

The J/S reduces as the square of the reducing distance, and the range at which the radar skin return is adequate to establish track in the presence of jamming is still called the burn-through range.

5

Characterization of EW Equipment

In order to make a model of a piece of equipment and to subsequently develop a simulation based on that model, it is necessary to first understand how the equipment actually works, then to understand the relationships between the controls, inputs, and outputs of that equipment. In this chapter, we will look at antennas, transmitters, receivers, processors, and direction-finding subsystems. In each case, we will first discuss the roll of that element in EW, and the different types or approaches used in EW. Then we will discuss the input requirements, operator or automatic controls, and output from that element. The object is to end with what is required to model the performance of that element.

5.1 Antennas

An antenna is characterized by both gain and directivity. If the receiving antenna is pointed in the direction of an emitter, the signal received from that emitter is increased by the antenna gain (which varies from about −20 dB to +55 dB, depending on the type and size of the antenna and the signal frequency). The antenna's directivity is provided by its gain pattern.

There are two basic classes of antennas used in EW and threat systems; one type covers 360 degrees of azimuth and the second covers a smaller angular area. Sometimes the 360-degree antennas are called *omnidirectional* antennas, but the name is misleading. A true omnidirectional antenna, or isotropic antenna, would transmit energy or receive signals equally over a full sphere. There are no truly isotropic antennas because of distortion from support structures and so on, but the concept is useful to us because we define its gain as unity (or 0 dB) and define the gain of all types of antennas relative to the gain of that imaginary isotropic antenna.

5.1.1 Antenna Pattern Definitions

The gain pattern shows the antenna's relative gain as a function of the angle from bore sight to the direction of arrival of the signal. The gain pattern is determined by placing the antenna in an anechoic chamber and rotating it as a calibrated signal is transmitted to the antenna. A receiver attached to the antenna records the received signal strength over a full rotation. The gain patterns are usually taken for 360 degrees of rotation on two mutually perpendicular axes (for example, horizontal and vertical). Figure 5.1 shows a pattern in one plane, and identifies important features of the gain pattern for a directional antenna; 360-degree antennas have fairly uniform horizontal patterns, but have vertical patterns somewhat like the figure.

This diagram shows antenna gain in polar coordinates. The large part of the gain pattern is called the *main beam* or *main lobe*. The antenna is designed for the characteristics of the main beam. The other lobes in the gain pattern (including the back lobe) are called *side lobes*.

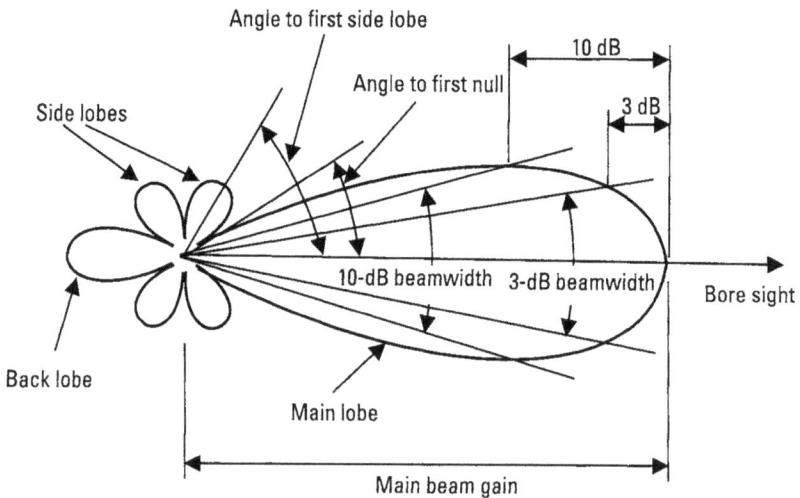

Figure 5.1 Antenna gain pattern with definitions.

The bore sight of the antenna is generally taken as the maximum gain point of the main beam. This is the direction the antenna is supposed to point. In terms of the discussion in Chapter 4, optimum link performance is achieved when the bore sight is pointed at the receiver (or the transmitter) or when the radar antenna bore sight is pointed at the target.

There is generally an antenna-pointing-accuracy specification in any program. This is the maximum angle that the bore sight can be misdirected. This pointing error reduces the antenna gain in the link or radar range equation. Note that this whole pattern approximates a sine *x/x* curve, where *x* is the angle from the bore sight, but actual antenna gain patterns are measured and published by antenna manufacturers.

The main beam directivity is usually defined by the antenna beam width. Beam width typically refers to the 3-dB beam width. The 3-dB point is the angle from the bore sight that the antenna gain is down to half of its peak value (i.e., 3 dB below peak gain). The 3-dB beam width is the angle (double sided) between the two 3-dB points. The 10-dB beam width is likewise defined as the (two-sided) angular distance between the two offset angles (in a single plane) between the two points at which the antenna gain is down to 10% of the peak value. When considering interference conditions, it is also sometimes useful to know the angle to the first null (i.e., the gain minimum between the main beam and the first side lobe) and the angle from bore sight to the peak of the first side lobe.

Side lobes are undesirable from the antenna designer's point of view. One of the quality factors of a directional antenna is how far down its side lobes are (in gain) from the gain at the peak of the main beam. In EW, we often do not know the side lobe level—or we are designing a system to operate against a large number of threats with different antennas. Therefore, we often use the assumption that there are zero dB side lobes. This means that the side lobes are as far below the main beam gain as the numerical value (in dBi) of the main beam gain. Another way to put this is that the side lobes are assumed to have the gain of an isotropic antenna.

5.1.2 Polarization

Antenna polarization is a characteristic of the type of antenna and its orientation. An antenna can be linearly polarized (vertical, horizontal, or any angle in between), or it can be circularly polarized (right- or left-hand circular).

If both the transmitting and receiving antenna in a link have the same polarization, there is no polarization loss. If they have different polarization, there will be a loss that decreases the level of the received signal, as shown in Figure 5.2.

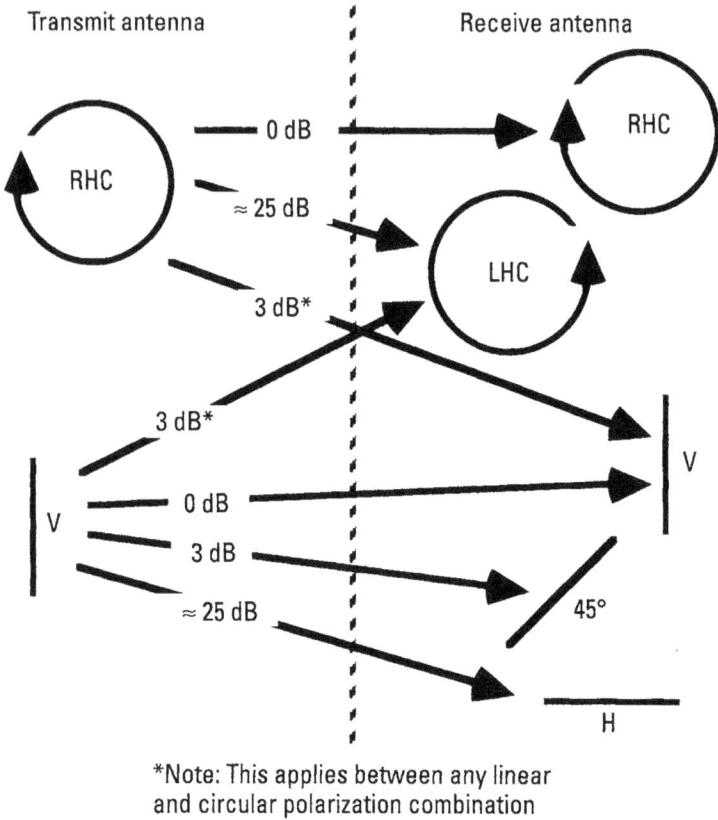

Figure 5.2 Antenna polarization losses.

You will note that the vertical-to-horizontal polarization mismatch can cause about 25 dB of polarization loss. This assumes that there are no significant cross-polarization components caused by the antenna mounting or the environment. The right-hand to left-hand circular polarization loss is shown as approximately 25 dB. This is very much an approximation, since small spiral antennas can have as little as 10-dB cross-polarization loss and spacecraft-to-ground-link antennas, which are carefully designed for polarization isolation, can have as much as 33-dB polarization loss.

An interesting item on this chart is the 3-dB loss between either circular polarization and any linear polarization. This is a trick widely used in EW to avoid large polarization mismatches when the polarization of threat transmitters cannot be predicated.

5.1.3   Types of Antennas

There are many types of antennas, but our discussion here is limited to a few typical antennas of importance in EW applications. The first antennas considered are those that provide 360-degree azimuth coverage.

Whip Antenna

The whip antenna shown in Figure 5.3 is common in tactical communication transceivers—it is therefore the primary threat antenna for communications electronic support and COMINT systems. The whip is an ideal antenna for vehicle or man-pack applications. As shown, over a ground plane, its vertical pattern is a slightly elevated lobe, which provides fairly constant gain of about 2 dBi to all azimuths for narrow-frequency-range coverage (down to 0 dBi for wide frequency ranges). Whips are common in HF, VHF, and UHF bands. They are vertically polarized.

When estimating the height of an antenna, the effective radiating center of a whip antenna should be taken as about one-quarter of the way up from the base of the whip.

Dipole Antenna

The dipole antenna (Figure 5.4) is used in many applications important to EW. They are often used as elements of other types of antennas or antenna arrays. Receiving antennas on aircraft are often folded dipoles. The dipole pattern is rather like a doughnut centered on the antenna. There are nulls at the end of the antenna and a fat lobe with its maximum gain perpendicular to the antenna and significant coverage ±30 to 45 degrees from the perpendicular. The peak gain of the dipole is as much as 3 dBi for narrow-frequency coverage, and may be as low as −15 to −20 dBi if it is used over several octaves. The polarization of a dipole is linear, aligned with the antenna. Dipoles are used from HF through microwave frequencies.

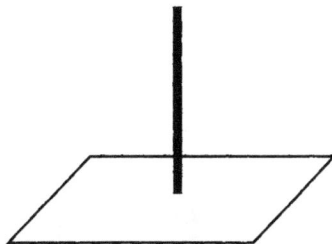

Figure 5.3 Whip antenna.

Figure 5.4 Dipole.

Yagi Antenna

The yagi antenna is a dipole arrayed with other elements to form a beam. A horizontal yagi as shown in Figure 5.5 will be horizontally polarized. It has a gain pattern that is narrower in the azimuth plane than in elevation, with on the order of 90 degrees vertical and 50 degrees beam width. It is usually used in the VHF and UHF frequency ranges. Its peak gain can be expected to be 5 to 15 dB and to cover about a 5% frequency bandwidth.

Log-Periodic Antenna

A similar type of antenna is the log periodic. It is an array of dipoles with their length and spacing designed to increase the frequency range. Log-periodic antennas have beam widths similar to yagis, but only about 6 to 8 dBi of gain. However, they can cover 10 to 1 bandwidths. Like the yagi, this antenna is linearly polarized in the plane of the dipole elements. These antennas are used from HF through microwave. Ground-based intercept and jamming systems that require antenna gain and wide frequency coverage often use log-periodic antennas which can be rotated around the axis of the antenna for polarization selection and their mounting masts rotated to targets of interest.

Figure 5.5 Yagi.

Direction-Finding Arrays

Direction-finding (DF) systems require arrays of antennas. These are often arrays of vertical dipoles. Figure 5.6 shows a dipole array configuration used for several types of DF systems. There can be as few as 3 dipoles in the array and as many as 12 (or more) depending on the DF technique and the required accuracy. Each dipole has the standard dipole gain pattern, although somewhat distorted by the presence of the support structure and the other dipoles. The dipoles in these arrays often have very low gain, because they are used over an octave (or greater) frequency range.

There are sometimes "Christmas tree" arrays of two or three of the pictured arrays on the same mast. The antennas and the length of the mounting arms are shorter for each level. This configuration extends the DF system's effective frequency range.

Cavity-Backed Spiral Antenna

Cavity-backed spiral antennas are widely used in radar warning receiver (RWR) systems because they are small and cover wide frequency ranges. A common type shown in Figure 5.7 covers 2 to 18 GHz in a single antenna. These antennas are also used in applications from VHF to millimeter wave.

Figure 5.6 VHF or UHF DF array.

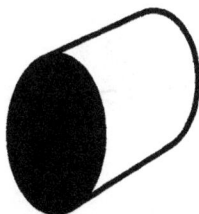

Figure 5.7 Cavity-backed spiral.

The peak gain of this type of antenna varies greatly with frequency. For example, the 2–18-GHz antenna has about 2-dBi gain at the highest frequency and about −12 dBi at the lowest. The gain pattern has approximately a cosine^2 variation. This makes it look like a ball sitting on the face of the antenna. Modern RWR antennas are designed to have a linear gain variation (in dB) versus angle from the bore sight, for all frequencies in range and out to ±90 degrees. Cavity-backed spirals are circularly polarized with the sense determined by the spiral.

There are new antennas that have the same exterior shape, but contain phased dipoles. These antennas can be switched to either right- or left-hand circular polarization.

Four cavity-backed antennas placed at about every 90 degrees around an aircraft provide an almost even gain coverage over 360 degrees in the aircraft's yaw plane.

Horn Antenna

A horn antenna is a flared waveguide that generates a well-behaved beam. A typical horn is shown in Figure 5.8. It is used from VHF through millimeter wave and usually provides 5 to 10 dB of gain. Typical beam width would be 40 degrees × 40 degrees. Horns have linear polarization, but the throat can be covered by a polarizer that converts it to circular polarization. Horn antennas can have about 60% efficiency (i.e., the percentage of the power into the antenna that is radiated within the 3-dB beam width.)

Phased-Array Antenna

A phased-array antenna is an array of relatively wide angular coverage antennas that are placed approximately one-half wavelength apart, as shown in Figure 5.9. The phase-delay network modifies the phases of the antenna outputs so that the signals add constructively (i.e., in phase with each other) when signals are arriving at the array from some specific direction. The figure shows a linear array, which provides directivity in one plane. If the array is planar, the directivity is in two planes. One type of phased-array antenna is a

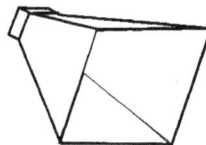

Figure 5.8 Horn antenna.

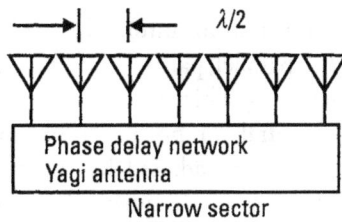

$\lambda/2$

Phase delay network
Yagi antenna

Narrow sector

Figure 5.9 Phased array.

flat plate with fixed delay lines to form a fixed beam at a fixed angle from the plate surface to the plate. The direction of the beam is changed by rotating the plate. Another type has fixed antennas, but can move the direction of the beam by coordinated adjustment of the individual phase delays in the network. Electronic steering (using electronically variable phase delays) allows the antenna to be instantly moved from one angle to another. A mechanically steered antenna must, of course, move linearly from angle to angle.

The antennas in the array are often dipoles, horns, or cavity-backed spirals. The polarization of the array is the polarization of the comprising antennas. The efficiency of a phased-array antenna is generally about 30%.

Many modern radars use phased arrays. Fighter aircraft sometimes use steered flat-plate antennas with fixed phased-array beams. Many new air-defense radars use electronic steering in elevation with mechanical azimuth steering.

Parabolic Dish Antenna

Perhaps the parabolic dish (see Figure 5.10) is the most important antenna in EW because it is used for most radars. The dish itself is a parabolic section

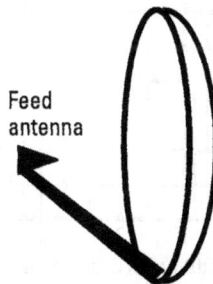

Feed
antenna

Figure 5.10 Parabolic dish.

(since a parabola is, by definition, infinite). The parabolic curve reflects rays from its focus to straight, parallel paths. The parabolic antenna has some kind of feed antenna at its focus. This can be almost any kind of antenna but is usually a log periodic, spiral, or horn. Optimum dish antenna design requires that 90% of the energy radiated by the feed antenna strikes the reflector. The efficiency of a narrow-frequency band dish antenna (covering up to about 10% bandwidth) can be up to 55%. Wider band antennas will have less efficiency—for example, a 2–18-GHz dish antenna can be expected to have about 30% efficiency.

5.1.4   Relationship Between Antenna Gain and Beam Width

The chart in Figure 5.11 shows the gain versus 3-dB beam width of a 55% efficient antenna. The value of this chart is to show at a glance the maximum performance that can be expected from an antenna. Since a radar operates at a single frequency or over a narrow range of frequencies, you would ordinarily assume that a threat radar has 55% efficiency unless you know otherwise.

Figure 5.12 is a nomograph that allows you to determine the gain of an antenna with a reflector of a given size at a specified frequency. Note that this assumes a symmetrical antenna (i.e., a round dish) and 55% efficiency. If the efficiency is different, the gain is adjusted as shown in Table 5.1.

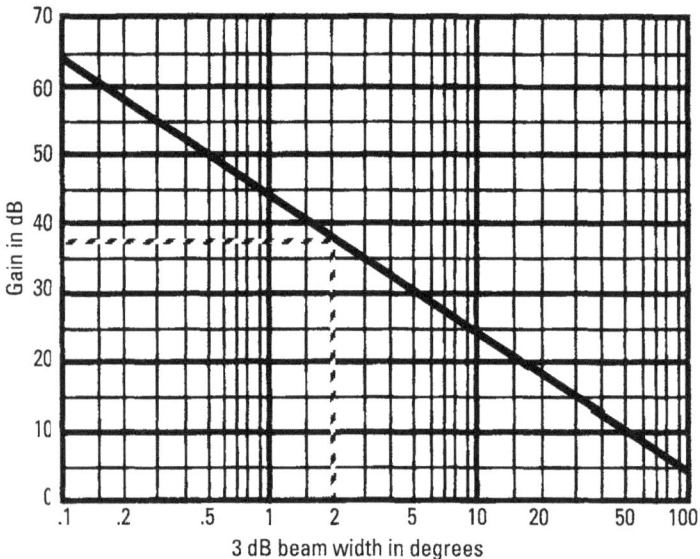

Figure 5.11 Gain versus beam width for 55% efficient antenna.

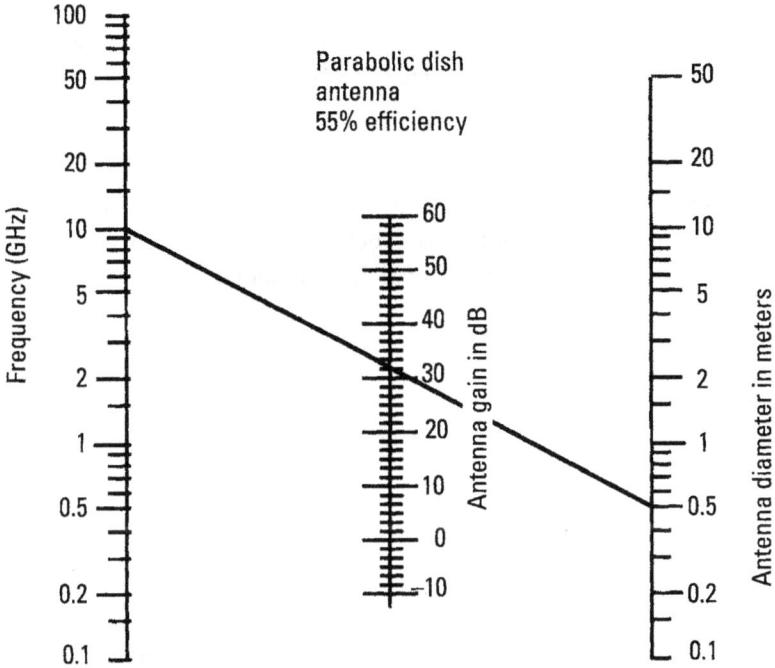

Figure 5.12 Parabolic dish antenna gain versus diameter versus frequency.

Table 5.1
Gain Adjustments Versus Efficiency

Antenna Efficiency (%)	Adjustment to Gain in Figure 5.12 (dB)
60	+0.4
50	−0.4
45	−0.9
40	−1.4
35	−2
30	−2.6

Several antenna companies have antenna slide-rules that allow you to directly calculate the trade-offs among size, frequency, efficiency, and gain.

Such a device can be obtained (for free) by contacting the marketing department at Tecom Industries (9324 Topanga Canyon Blvd., Chatsworth, CA 91311-5795; (818) 341-4010).

Nonsymmetrical Antenna Gain Versus Beam Width

If the antenna reflector is not symmetrical, the resulting beam will be taller if the reflector is wider and vice versa. There are convenient formulas for the gain of such an antenna. For a 55%-efficient parabolic dish

$$\text{Gain (not in dB)} = 29{,}000 \,/\, (\theta_1 \times \theta_2)$$

where θ_1 and θ_2 are two orthogonal 3-dB beam widths (in degrees). For a 60%-efficient horn antenna the formula is

$$\text{Gain (not in dB)} = 31{,}000 \,/\, (\theta_1 \times \theta_2)$$

5.1.5   Determining Antenna Size

In an analysis, you may need to estimate the size of an antenna in order to predict its beam width and gain for an EW engagement model. While it is sometimes difficult to learn the size of a threat antenna, the diameter of the missile or the nose of the aircraft it is mounted in is readily available. There are a couple of rules of thumb that you will find useful in converting a vehicle diameter into an antenna diameter. If the antenna is a mechanically steered dish, assume that it is 85% of the diameter of the vehicle. If it is a mechanically steered phased array, assume that it is 90% of the diameter of the vehicle. If it is an electronically steered phased array, assume 95%.

5.1.6   Modeling of Antenna Performance

It is common practice to model the effect of an antenna by adding its peak gain and adjusting for the gain reduction from the gain pattern. The following discussion assumes that all gains are in dB and all signal strengths are in dBm.

A 360-degree communication type antenna has a gain pattern that is essentially constant as seen by all of the other antennas that will interact with it. For a transmitting antenna, simply add the antenna gain to the transmitter power to determine the ERP in the direction of the receiver. For a receiving antenna, add the antenna gain to the power arriving at the receiver location (P_A) to determine the received power.

If there are directional antennas involved, it is a little more complicated, as shown in Figure 5.13. Start with the peak gain of the main lobe of

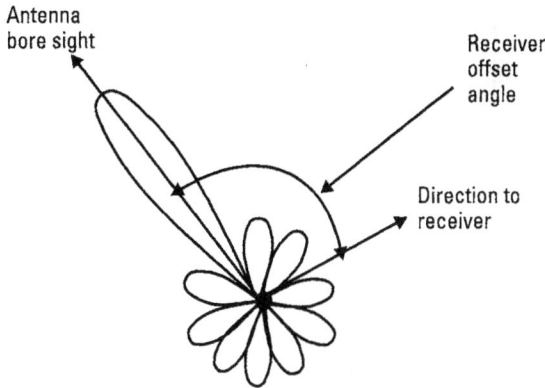

Figure 5.13 Modeling of transmitting antenna.

the antenna, then subtract the difference between the off axis gain at any given offset angle from that peak. For a transmit antenna, determine the angle between the antenna bore sight and the receiver location. Determine the gain as above for that offset angle, and add that to the transmitter power to calculate the ERP (in the direction of the receiver). For a receiving antenna, determine the angle between the antenna bore sight and the transmitter location. Determine the gain reduction at that offset angle and subtract it from the bore sight gain. Then add the remaining gain to the P_A to determine the received power.

5.2 Transmitters

The qualities of the transmitter important to EW modeling and simulation are as follows:

- The location of the transmitter (transmitted signals start at this location);
- Transmitted power (which is added [in dB] to the antenna gain to constitute the effective radiated power [ERP]);
- Signal RF frequency;
- Signal modulation.

The throughput function of the transmitter is as shown in Figure 5.14. While the internal details of the transmitter may be different from the

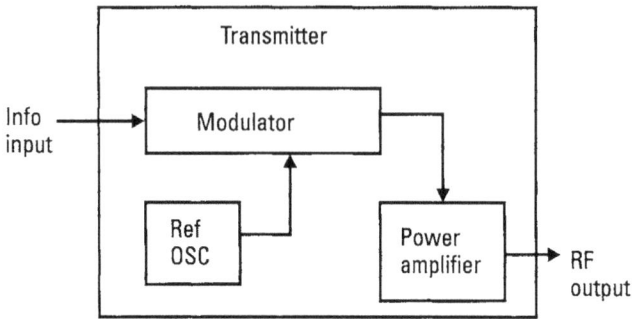

Figure 5.14 Transmitter functional diagram.

arrangement shown in the figure, this is what the transmitter does. The information signal is modulated on to a carrier generated by the RF oscillator. The modulated signal is amplified to the transmitted signal level.

The modulations considered in EW are pulse, continuous AM, continuous FM, FM on pulse (chirp), binary phase shift on pulse, or one of the modulations associated with digital data.

The transmitter in an EW model usually represents a threat signal, so its location will represent the location of the threat transmitter. Threat locations will often be grouped; for example, AAA radars and SAM radars protecting some enemy asset. Communication transmitters may be located close to the radars in an integrated air-defense network.

Since the threat may move in response to movement of some friendly asset (e.g., our aircraft or ship), the location of the threat transmitter may be a function of the friendly asset location. For example, an enemy fighter (AI) may make a frontal or rear-firing pass at a friendly aircraft.

A radar threat signal will generally change modes in response to the location of friendly assets, as shown in Figure 5.15. For example, a fire-control radar will go from acquisition mode to tracking mode when the friendly asset is within tracking range, and into a launch mode when the friendly asset is within the lethal range of the weapon. Note that this mode change may involve a change in the scan characteristics of the antenna associated with the radar as well as modulation changes or the initiation of the signal.

Another example of responsive transmitter movement is an antiship missile homing on a ship. The transmitter will turn on about 10 km from the ship and approach the moving ship's changing location at the missile velocity. If a ship-protection countermeasure is used, it will cause the missile to move toward a position away from the ship—following the missile guidance rules in the model.

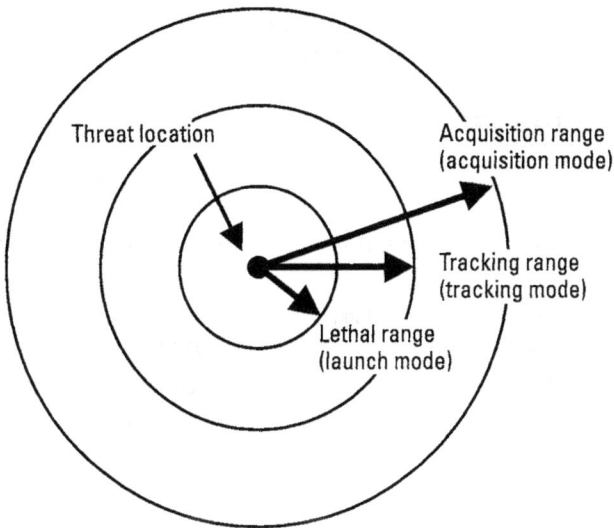

Figure 5.15 Threat modes versus friendly asset location.

5.3 Receivers

This section discusses the types of receivers used in EW systems, their general operating characteristics, their sensitivity, their control mechanisms, and the ways they are characterized in EW models.

5.3.1 Types of Receivers

EW receivers are divided into wideband and narrowband types. There are several important wide band types: crystal video, IFM, compressive, channelized, and digital. The important narrowband types are fixed tuned and superheterodyne. Table 5.2 summarizes the characteristics of the receiver types. The following discussions characterize the throughput function of each receiver type in preparation for later discussions of receiver simulators (for example, "RF signal to digital frequency readout").

Crystal Video Receiver

The crystal video receiver (CVR) is used primarily as a pulse receiver, and primarily in the microwave range (although it has been used down to VHF). It is used in almost every radar warning receiver.

As shown in Figure 5.16, the CVR accepts a wide frequency range and directly detects the amplitude modulation of any signal present and amplifies

Table 5.2
Types of EW Receivers

Receiver Type	Frequency Range	Sensitivity	Unique Characteristics
Crystal Video	VHF to μw	Low	Receives all pulses in frequency range. Can receive only one signal at a time
IFM	μw	Low	Determines only frequency of signal. Can receive only one signal at a time.
Compressive	VHF to μw	High	Determines only frequency of signal. Can receive simultaneous signals.
Channelized	HF to μw	High	Recovers modulation. Can receive simultaneous signals.
Digital	Any	High	Software controlled. Can perform virtually any receiving function.
Fixed tuned	Any	High	Recovers modulation. Can only receive signal at one frequency.
Superhet	Any	High	Recovers modulation. Can isolate one signal from environment.

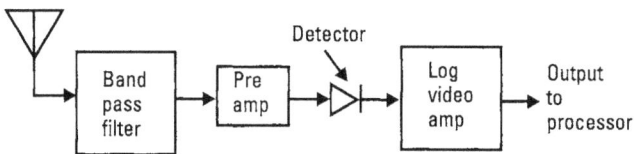

Figure 5.16 Crystal video receiver.

the resulting video in a log video amplifier. The video output contains the modulation of every signal present, so the receiver works best in a pulse environment in which each signal has a low duty cycle. When pulses from different signals overlap, the individual pulses cannot be clearly received for subsequent processing. When there is a strong CW signal anywhere in the covered band, the output of the CVR is significantly reduced.

The sensitivity of modern CVRs is approximately –40 dBm. With the use of an appropriate preamplifier, its sensitivity can be improved to about –65 dBm. This is adequate to receive the main beam of radars beyond the range at which the radars can detect typical aircraft, but is usually not adequate to receive radar side lobe transmissions.

The throughput of the CVR is RF to video. The CVR output will contain the pulses of any transmitter that is close enough to provide signal strength above the sensitivity level. The video output will be proportional to the log of the received RF signal strength.

When a strong CW signal is present at the same order of magnitude as pulse signals being received, the output pulses are reduced, and their received signal strength cannot be dependably measured. Because overlapping pulses have the same effect, the high PRF of a pulse Doppler radar anywhere in the band will, in effect, make a standard CVR output useless.

Instantaneous Frequency-Measurement Receiver

The instantaneous frequency measurement (IFM) receiver can determine the frequency of a received signal in a very short period (about 50 ns).

As shown in Figure 5.17, the IFM requires a hard limiting amplifier. This is because the output of the IFM circuitry varies both with signal strength and frequency. In this configuration, the IFM outputs a digital representation of the frequency of any signal present at its input. The IFM must operate on a single signal during its very short measurement period. If more than one signal is present, the output is nonsense; however, it is possible for a processor to determine that the IFM is not generating valid frequency data.

Like the CVR, the IFM is made useless with the presence of a single CW signal anywhere in the band it covers. However, its great speed allows it to handle overlapping pulses by measuring the frequency during the parts of the pulse not overlapped.

The sensitivity of the IFM is approximately the same as that of the CVR. It can measure frequency to approximately 0.1% of the frequency range it covers.

Compressive Receiver

The compressive receiver, also called the *microscan receiver,* measures the frequency of every signal present in a wide frequency range. It has better sensitivity than the IFM, and unlike the IFM it can handle multiple simultaneous signals.

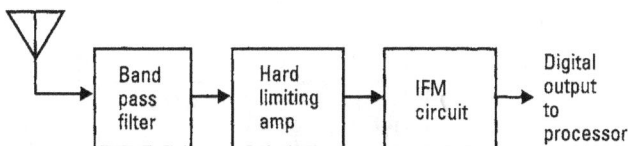

Figure 5.17 IFM receiver.

The principle of operation of the compressive receiver involves sweeping a local oscillator (LO) at a very high rate of speed—too fast to allow detection of any signals encountered. However, as shown in Figure 5.18, the detector is downstream of a compressive filter. This filter causes a delay in the output that is proportional to frequency. The slope of this delay versus frequency matches the sweep rate of the LO, so signals, in effect, pause in the detector long enough to be detected. While this process precludes recovery of the modulation, it allows measurement of the RF frequency of all signals in the swept band. A powerful use of the compressive receiver is to allow a narrowband receiver to be rapidly set to the frequency of an important threat signal at an unknown frequency.

The throughput of this receiver is RF signal present to digital frequency output. That is, any time a signal is within the swept frequency range, a digital representation of its frequency will be present in the list frequencies output by the receiver.

Channelized Receiver

A channelized receiver comprises a set of fixed tuned receivers that cover adjacent portions of a frequency range. As shown in Figure 5.19, the input signal is multiplexed into a bank of receivers. It is common practice to design the filters to the individual receivers so that adjacent filter pass characteristics cross at the 3-dB point.

Since each of the individual receivers in the channelized receiver can have full demodulation capability, the throughput is RF to demodulated audio or video (any type of modulation for which the channel receivers are designed). A channel will output the proper modulation any time a signal with the appropriate modulation type is present at the frequency it covers.

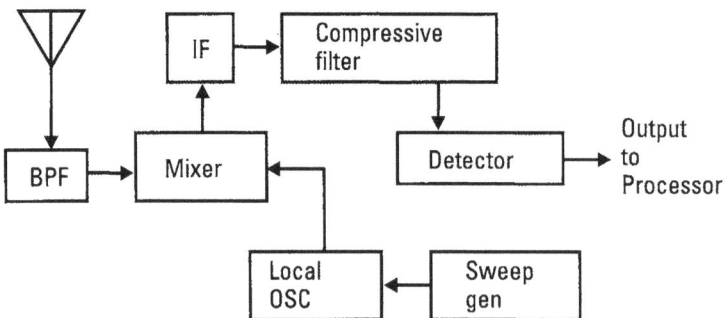

Figure 5.18 Compressive receiver.

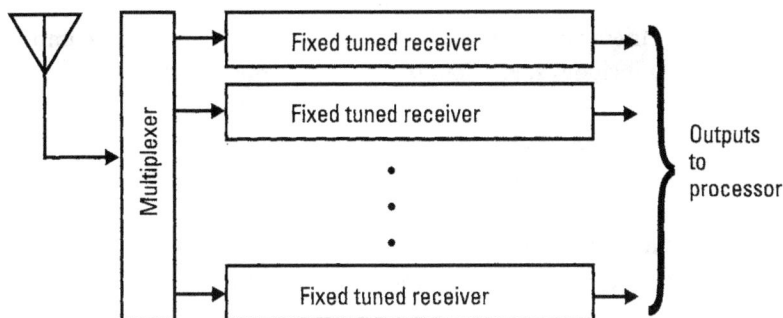

Figure 5.19 Channelized receiver.

Digital Receiver

The future of EW seems to belong to the digital receiver, because of its ability to cover a relatively wide frequency range while performing a variety of detection and analysis functions. The digital receiver is in effect a digitizer and a computer, as shown in Figure 5.20. The digitizer converts a frequency range into digital words and the computer performs all of the normal receiver function in software. The software functions also include some analysis techniques that are difficult to implement in hardware.

Earlier digital receivers digitized a zero IF, which was an intermediate frequency band that was translated down in frequency until its lower edge was at a very low frequency. With the availability of more capable analog-to-digital converters (ADC), it was possible to digitize the actual IF frequency, simplifying receiver design and improving output quality.

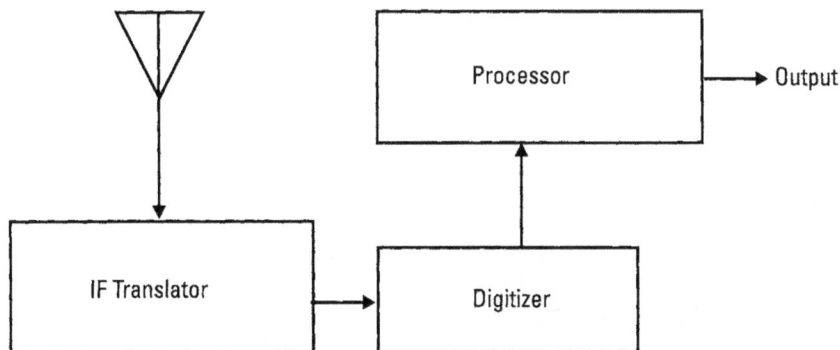

Figure 5.20 Digital receiver.

A fast Fourier transform (FFT) function is performed in the computer, which typically includes specialized digital processors to perform this type of high-throughput function. The output of the FFT is a digital spectrum representation of the input range. The frequency resolution elements are called *channels*. If there are two digitization inputs in phase quadrature, the computer can determine the phase of signals in each channel. (This will be seen to be useful in direction-finding schemes, discussed in Section 5.5.) The computer can also determine the modulation in each FFT channel, and demodulate the signals.

Because of its great flexibility, the throughput function of the digital receiver can be almost anything. However, in EW applications it will normally be characterized by RF inputs to signal modulation and digital frequency information. Any time a signal is in the range covered by the digital receiver, the output will have the frequencies of all signals present, and might also have the demodulated output from one channel, depending on operator-instituted or automatic mode commands.

Fixed Tuned Receiver

Fixed tuned receivers are useful for monitoring important, continuing functions. One example is a time or weather channel receiver. A more interesting application is in Global Positioning System (GPS) receivers. They are designed to receive signals from different satellites—all at the same frequency, but modulated with different digital codes that spread the signal spectrum. When the correct code is applied in the receiver, only a single signal is received. All others are suppressed.

The throughput of a fixed tuned receiver is signal at some specific frequency to received signal modulation. Any time a signal with appropriate modulation is present at the frequency to which the receiver is tuned, the modulation on that signal will be present in the receiver output.

The output of a GPS receiver will usually be represented by high accuracy time and location information anytime the receiver is located so as to have line of sight to a large angle of the sky.

Superheterodyne Receiver

A superheterodyne, or superhet, receiver provides high sensitivity and can be used against any type of signal. It is widely used in communication-band intercept and emitter-location systems and in microwave naval ESM systems. It is also used in modern RWRs to handle CW and pulse Doppler signals.

A superhet receiver covers a narrow frequency range with relatively high sensitivity (the sensitivity is proportional to the bandwidth covered). As shown in Figure 5.21, the superhet usually has a tuned preselection filter that

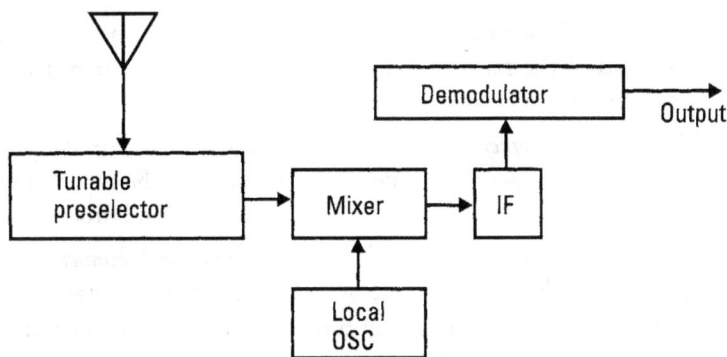

Figure 5.21 Superheterodyne receiver.

isolates a segment of the input frequency range. Then, a local oscillator converts the selected segment to a fixed intermediate frequency (IF) which is amplified and filtered in the IF amplifier (called simply IF on the diagram; a commonly used shorthand). Note that the process of translating a frequency band to another RF frequency is called *heterodyning.*

The throughput of a superhet receiver is RF at the frequency to which the receiver is tuned to modulation on the received signal. Any time the tuning commands (manually or automatically generated) place the superhet receiver at the frequency of a signal present in the tuning range, and the correct demodulation has been selected, the demodulated information will appear at the receiver output. Some superhet receivers have selectable IF bandwidths; a narrower output normally reduces output noise (i.e., causes higher output SNR).

5.3.2   Receiver Subsystems

The above described types of receivers are often used in combination. The following are common configurations:

- *Multiple CVRs, each covering a portion of the radar threat frequency range.* This allows a large frequency range (e.g., 2 to 18 GHz) to be covered in bands (e.g., four bands of 4 GHz).

- *Multiple CVRs and an IFM receiver.* The IFM follows a switchable frequency translator so that it can cover any of the CVR bands. The four bands could also be folded into a single frequency band—with all frequencies present at the IFM input. The IFM would measure the frequency of any signal (pulse) in its input band and the CVRs (which are in separate bands) would resolve (by time coincidence) the band of the measured pulse. A processor could then determine the actual RF frequency of each measured signal.

- *Compressive receiver with several superhet receivers.* The compressive receiver determines the frequencies of signals and the superhet receivers are set to those frequencies for signal intercept and processing.

- *Several communication-band superhet monitor receivers with a direction-finding (DF) subsystem that includes one or more superhets.* The DF system automatically tunes to the frequency of a monitor receiver and measures direction of arrival when requested to do so by the operator.

5.4   Processors

Processors perform a wide range of functions in EW systems, many of which are transparent to operators. They control system assets, de-interleave signals, perform signal identification, run emitter-location algorithms, read and implement operator commands, and drive displays.

In general, the way that processors are modeled is to assume that their software has worked, and simply create (by computer) the results that the EW system processor would have produced with the signal and command inputs present in the model or simulation.

5.4.1   Signal Identification

Signal identification generally involves the following:

- Storage of a set of parameters for all considered signal types in a table;
- Comparison of the parameters of collected signals with the values in the table;
- Reporting of the applicable signal type when a parametric match to a signal type in the table is found;
- Reporting an unknown signal if no match is found.

This process gets somewhat more complex when there are ambiguities among signal types. When modeling such an analysis situation, it is necessary to have rules for how to resolve (or not resolve) the ambiguity.

Another exciting aspect of processor modeling is the representation of performance anomalies in specific EW systems. The model must include rules for recognizing situations in which the anomalies will occur (i.e., high G turns, environment density above some specific level, the simultaneous presence of two or more specific signal types in some geometric relationship, etc.). Then, the processing outputs must be appropriately generated.

5.4.2 Operator Interfaces

Operator interfaces on modern EW systems are largely computer controlled, so there is little challenge in making the model or the simulation match the real systems being modeled or simulated. However, when analog controls are present, the fidelity with which analog functions are read can become an issue. These aspects, along with update rates and so on, will be covered in detail in Chapter 10.

5.5 Emitter Location

Emitter location is an important aspect of ES. Knowing the location of emitters allows the development of electronic order of battle, evaluation of the immediacy of threats, development of targeting data, and handoff of threat-location information to other systems and platforms.

Direction finding (DF), as opposed to emitter location, can also be a valuable EW tool. Just knowing the direction of arrival of signals can often allow the determination that there is more than one emitter operating in an area. It can be valuable in the deinterleaving of pulses into individual threats. Also, it allows homing on an emitter location.

This section describes the important DF and emitter-location techniques used in EW. The basic theory for each approach is explained, and its characteristic accuracy and applications are described. Finally, the implications of each technique in the modeling and simulation of systems is discussed.

5.5.1 Basic Emitter-Location Approaches

These basic approaches are depicted in two dimensions, but can as well be applied in three dimensions.

Azimuths from Two Locations

Measurement of the azimuth of arrival of a single signal from two known locations allows the determination of the emitter location in a plane (see Figure 5.22). This process is called *triangulation*. The accuracy of emitter location depends on the following:

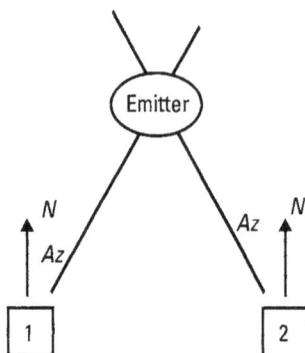

Figure 5.22 Triangulation emitter location.

- The accuracy with which the location of the DF systems is known. This part of the problem has become much easier since the availability of GPS receiver systems.
- The accuracy of the angular reference (usually North). This can be from a magnetometer (basically a digitally read compass), an inertial navigation system, or a manual survey.
- The accuracy of the angular measurement.

A system that determines emitter locations by triangulation typically uses three DF stations. These three stations develop three triangulation answers, which can be evaluated. The closer these three answers group together, the more accurate the location answer (usually the average of the three) is considered to be. Obstruction of the line-of-sight path between one of the DF stations and the emitter can cause grossly incorrect location calculations in a dense environment (i.e., the two stations might be looking at different emitters). Having three stations allows the detection of these types of problems.

Distances from Two Locations

The measurement of two distances from known locations determines the location of an unknown point in a plane, as shown in Figure 5.23. The emitter-location accuracy is determined by the accuracy of the location of the sensor sites and the accuracy of the distance-measurement technique used.

One Angle and One Distance from One Location

As shown in Figure 5.24, the location of an emitter in a plane can be derived from a single site if the angle of arrival from some reference and the distance

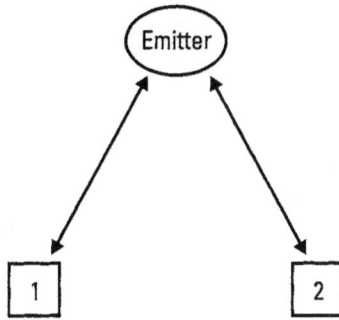

Figure 5.23 Emitter location with two distances.

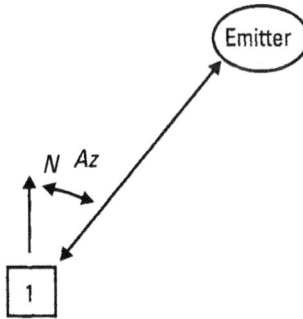

Figure 5.24 Emitter location with one angle and one distance.

to the emitter can be measured. This technique is commonly used in RWRs (with limited accuracy).

The emitter-location accuracy is determined by the accuracy of the location of the site, the angular-measurement accuracy, the angle-reference accuracy, and the accuracy with which distance is measured.

Two Angles from One Location

This technique is appropriate in two common applications: the single-site locator (SSL) and the location of ground emitters from an aircraft. Both techniques measure the horizontal angle of arrival conventionally and use a vertical angular measurement to determine the range to the emitter.

As shown in Figure 5.25, the SSL determines the distance to an HF emitter by measuring the elevation angle of arrival and determining the effective height of the ionosphere.

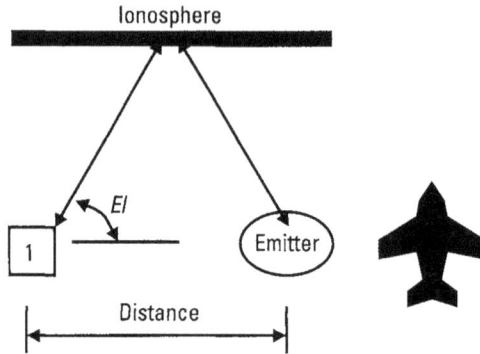

Figure 5.25 Distance in SSL measurement.

An aircraft with an inertial guidance package knows its own elevation and location. If the location system has a digital map of the area over which the aircraft is flying, it can determine the distance from the aircraft that the ray from the aircraft to the emitter intersects the ground (see Figure 5.26).

Location Accuracy

Emitter-location accuracy is typically described in terms of circular error probable (CEP). By definition, CEP is the diameter of a circle centered on the true emitter location into which half of the answers will fall. The CEP can be derived mathematically from the accuracy factors stated above for each of the basic emitter-location approaches. When the error area is not

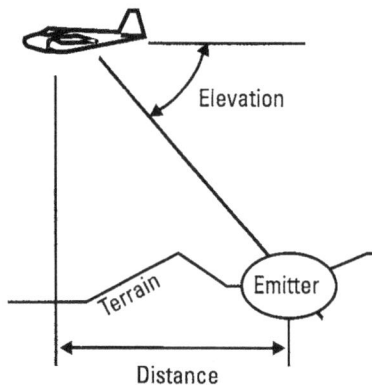

Figure 5.26 Distance from elevation angle and terrain.

symmetrical, an error ellipse is sometimes defined—along with an elliptical error probable (EEP).

The measurement accuracy is usually specified as a root mean square (RMS) accuracy. The RMS accuracy is determined by making many measurements over a range of conditions (e.g., at different angles and frequencies). Each error is squared, those square errors are averaged, and the square root of that average is the RMS error.

The following subsections describe actual techniques used to determine DF or emitter location.

5.5.2   Narrow-Beam Antenna

The technique used by radars to determine the angle to a target is also used by ES systems—particularly naval ESM systems. As shown in Figure 5.27, a sweeping narrow-beam antenna determines the angle of arrival by determining the pointing azimuth of the antenna when receiving a peak amplitude response from the emitter. Since the shape of the receiving antenna beam is known, multiple measurements across the scanning beam can be used to calculate the angle at which the peak response would have occurred. This technique is slow (requires antenna rotation) but can produce 1-degree or better accuracy if the antenna is large enough, and can measure each of several simultaneous signals.

5.5.3   RWR Amplitude Comparison

Figures 2.27 and 2.28 showed a basic RWR block diagram and the locations of the antennas on the aircraft. Figure 5.28 shows the patterns of those four

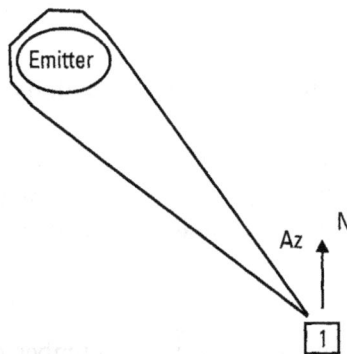

Figure 5.27 DF with narrow-beam antenna.

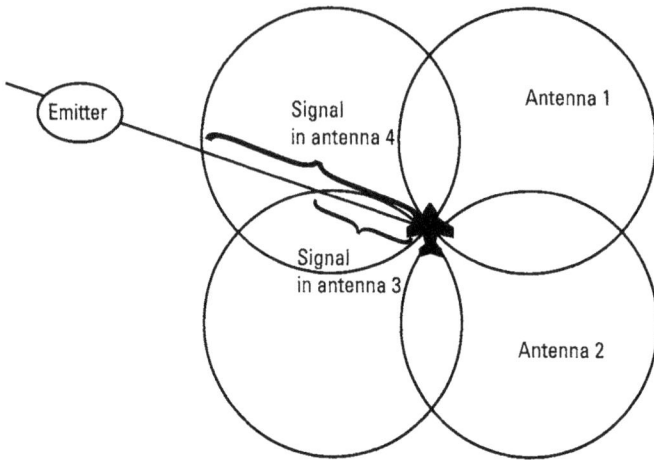

Figure 5.28 RWR antenna patterns.

antennas. By comparing the amplitude with which the antennas receive the signal, the system can calculate the angle of arrival relative to the nose of the aircraft. Taking a vector sum of the inputs from the two antennas closest to the angle of arrival allows the received signal strength of the signal to be determined.

Because the RWR processor identifies the type of emitter, the system knows the ERP of the radar in its main beam. With this information, the system can use the one-way link equation formula to calculate the approximate range to the emitter.

This approach provides monopulse DF, but is not highly accurate. The inaccuracy is due to reflections from the aircraft that reduce the accuracy of the antenna patterns—hence the accuracy of the amplitude comparison. The second cause of inaccuracy is the distance-measuring technique. First, the emitter ERP is not accurately known, and second, the propagation losses are impacted by many factors.

The generally stated accuracy for this kind of emitter location approach is about 15 degrees of angle and 25% of range. When higher accuracy is required, one of the other emitter-location techniques is used.

5.5.4 Watson-Watt

The Watson-Watt technique is used in a number of moderately priced communications-band DF systems. It uses an antenna array as shown in Figure 5.29. There can be any even number (four or more) of antennas

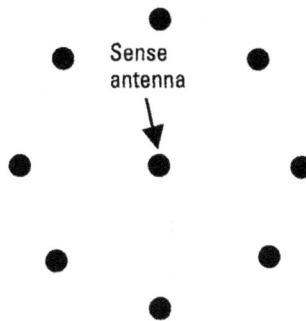

Figure 5.29 Watson-Watt antenna array.

around the outside. The DF system uses three receivers. One is attached to the center (reference) antenna, and the other two are switched to two of the outside antennas that are across the array from each other. By making sum and difference measurements, the three antennas form a cardioid pattern. The selected outside antenna pair rotate around the array, causing the cardioid to rotate. This allows the processor to determine the direction of arrival of the signal.

Although Watson-Watt DF systems can be calibrated to higher accuracy, most have RMS accuracy of about 2.5 degrees.

5.5.5 Doppler

The Doppler principle causes the frequency of a received signal to increase if the transmitter and receiver approach each other and to decrease if they move away from each other. By rotating one antenna around another, as shown in Figure 5.30, the frequency difference between moving antenna A and stationary antenna B is sinusoidal at the rotation frequency. The direction of arrival of the signal is at the azimuth, at which the sinusoidal frequency difference goes through the negative-going zero crossing.

Doppler DF arrays can be a circle of dipoles around a central reference antenna. The individual outside antennas are switched into one receiver in sequence, having the effect of rotating a single antenna around the reference antenna. There can be as few as three outside antennas.

5.5.6 Interferometer

When about 1-degree accuracy is required over an instantaneous angular segment (up to 360 degrees), the approach of choice is the interferometer

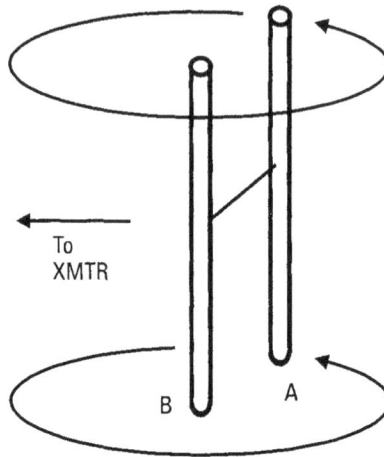

Figure 5.30 Principle of Doppler DF.

technique. This type of DF usually uses an array of irregularly spaced, linearly arrayed antennas or an array of vertical dipoles, as shown in Figure 5.31. This DF approach uses baselines—pairs of antennas—in which the phase is compared. In the linear array, the farthest-spaced antennas yield the best accuracy, but with ambiguities. The closest antennas resolve the ambiguities. In the vertical dipole array, every pair of antennas is used to make a total of six baselines.

The interferometric principle is based on the interferometric triangle, shown in Figure 5.32. The baseline is a pair of antennas at known locations, so their orientation and spacing is known. The baseline is the line in space connecting the two antennas. The wave front is a line perpendicular to the direction of arrival of the signal. It can be considered a line of constant RF

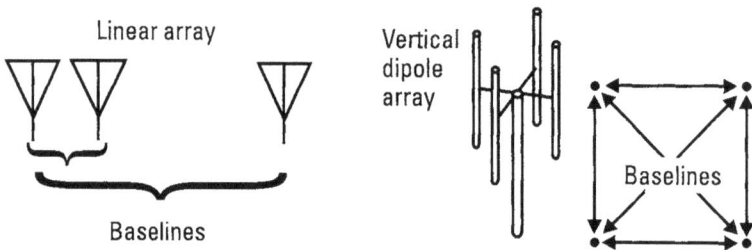

Figure 5.31 Antennas for interferometer DF.

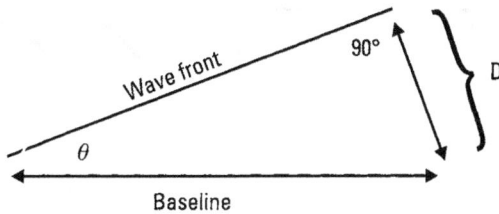

Figure 5.32 Interferometric triangle.

phase for the received signal. The distance D is determined by comparing the phase at the two antennas. If we measure the frequency of the signal, we can calculate its wavelength. The phase difference divided by 360 degrees and multiplied by the wavelength is the distance D.

Since we know both the baseline (the hypotenuse of the triangle) and D (the side opposite angle θ) we can determine θ. Note that the angle between the direction of arrival and the perpendicular to the baseline equals θ.

There is a front/back ambiguity that is resolved by use of directional antennas in the linear array or by measuring another baseline in the vertical dipole array to resolve the ambiguity.

5.5.7 Time Difference of Arrival

Time difference of arrival (TDOA) is a technique used for precision emitter location. We know that signals propagate at the speed of light, so if we knew the time the signal left the transmitter and the time it arrived at the receiver, we would know the distance. The problem, of course, is that we have no way to know when a noncooperative signal leaves the transmitter. Therefore, we measure the difference between its arrival time at two receivers.

If the signal is pulsed, the time difference can be determined by measuring the two arrival times against a very accurate time standard (like GPS) and subtracting. Signals with analog modulation are more challenging. They require that the signal be digitized at each receiver many times—each time with a differential delay. The two digital representations of the signal are correlated at each delay value. The delay value that peaks the correlation represents the TDOA.

Once the time difference is determined, the system knows that the emitter is on a curve. Since time difference is typically measured very accurately, the system is very close to that curve, but the curve is a hyperbola, which is infinite. Figure 5.33 shows the family of curves for different time-of-arrival differences. Each of these curves is called an *isochron*.

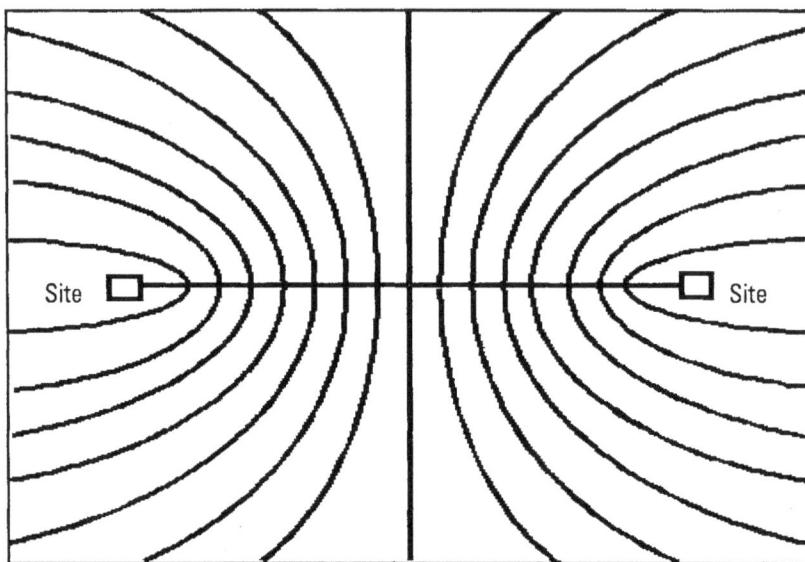

Figure 5.33 TDOA isochron contours.

If there is one more time-of-arrival measuring site, we can perform the TDOA algorithm with another pair of sites. That baseline will develop a second hyperbola that will cross the first one. The crossing of these two hyperbolas (isochrons) defines the location of the emitter to very great accuracy.

This technique works at any frequency range, and the time measurement sites can be either stationary or moving (including fast-moving aircraft).

5.5.8   Frequency Difference of Arrival

To understand frequency difference of arrival (FDOA), suppose that two aircraft measure the frequency of the signal received from a single emitter on the ground. The measured frequency at each aircraft will be changed as a function of the velocity of the aircraft multiplied by the cosine of the angle between the aircraft's velocity vector and the ray connecting the aircraft to the emitter, as shown in Figure 5.34. This will locate the emitter along a curving line connecting the two aircraft (it is called an *isofreq*). Each aircraft can have any velocity vector and any location—the computers can figure out the curves. However, the curves in Figure 5.34 are drawn for two aircraft traveling in the same direction at the same speed—purely to make the curve easier for us humans to understand.

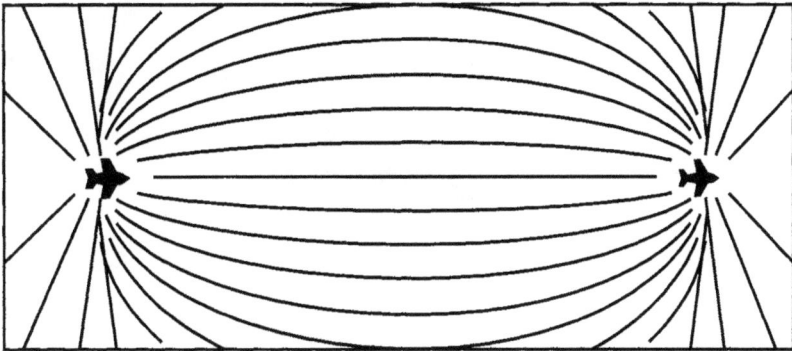

Figure 5.34 FDOA isofreq contours.

Again, we have located the emitter as being very accurately along a very long curve. If there is a third aircraft, we can draw another curve that crosses the first isofreq curve at the aircraft location

TDOA/FDOA

If TDOA and FDOA are implemented in the same two aircraft, the isochron and the isofreq lines will cross at the emitter location, so a full location can be accomplished with only two platforms.

6

Threat Modeling

Threat modeling is an exercise in determining, capturing, and reproducing the perception of a threat transmitter by a sensor. The point of view from which the threat must be considered is that of the sensors in EW systems. You need to look out at the world through the connector at the front end of a receiving system. What does it see and when does it see it? To make this work, of course, you need to consider what the threat is doing and why—in the momentary tactical situation. Then you need to consider what signals the threat is emitting and the direction in which it is emitting them, both as a function of time. Finally, you need to consider what that looks like (electronically) from the location of the receiver and what the receiving antenna does to arriving signals on their way to the receiver.

You may well need to consider how those threat signal look at some point deep into the EW receiving system: at an intermediate frequency (IF) or processor input or at an operator display. The way the signal looks inside the system will typically require that you consider the status of system controls, whether the control inputs come from a computer program or from an operator.

Modeling issues inside the EW system are dealt with in later chapters. In this chapter, we look at the modeling of threat signals on their way to the EW receiver.

6.1   Modes of Operation

One of the characteristics of a threat is its location. We also need to make some assumptions about how the threat reacts to various tactical situations. Normally, the proximity of a friendly asset (our friend, not the threat's friend) will cause the threat to change modes. Generally, a threat has the following types of operating modes:

- *Search,* in which a threat radar tries to determine the presence of potential targets;

- *Acquisition,* in which a threat radar detects the presence of a potential target and establishes a track (or lock) on the target;

- *Tracking,* in which a threat radar continues to update its tracking information while preparing to fire a weapon and while it guides a weapon to the target;

- *Launch,* in which an enemy weapon is fired;

- *Guidance,* in which the enemy sends guidance commands to a weapon in flight;

- *Fusing,* in which a short-range radar or other sensor determines exactly when to fire the warhead.

These steps often employ different radars, but many systems use the same radars (in different modes) to perform multiple functions.

6.1.1   Ground-Based Weapons

Missile systems on ships and at fixed or mobile ground sites will typically have the following important activities, each characterized by some range to the potential target. The actual ranges for each activity are derived from intelligence data; they will be a function of RCS of the target and some geometrical considerations (altitude, terrain, etc.). Figure 6.1 shows a typical set of mode division contours.

Search

Search is performed by long-range radars that operate continuously. Since their purpose is to detect new targets, they can be expected to be up and working when the EW system arrives over the horizon. They will continue their transmissions through the entire engagement.

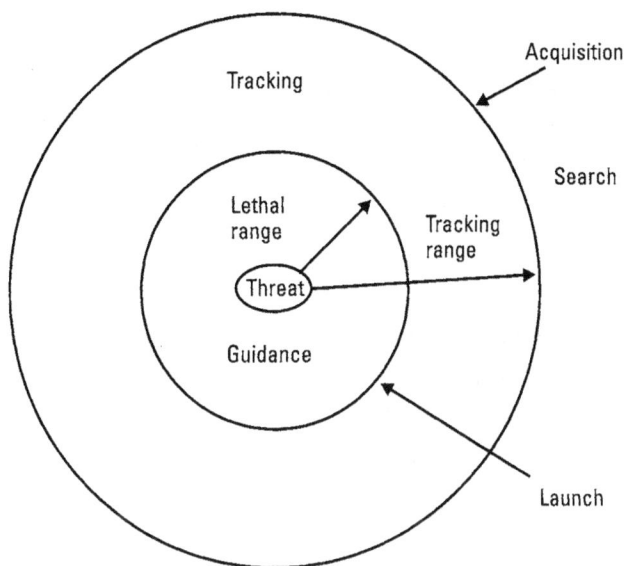

Figure 6.1 Zones for threat modes.

Acquisition

Acquisition is the event at which the mode of the radar relative to a given potential target changes from search to tracking. Sensors on that target will then observe that the tracking radar is tracking it.

Tracking

Tracking is performed by radars that have an effective range slightly beyond the lethal range of the weapons they support. Thus, when a friendly aircraft comes within the circular area of radius equal to a little over the lethal range (perhaps 10%), this radar can be expected to come on in a mode appropriate for tracking the target. Note that a tracking radar may be up when the potential target is beyond its lethal range, but the radar will be tracking another target that is within its tracking range.

Launch

Launch will occur at a range that is established in the model. You may chose to have launch occur when the target first comes within lethal range, or you may chose to launch at some proportion of lethal range (e.g., half of lethal range). You can run the model with different launch criteria to evaluate the relative weapon performance.

Guidance

Guidance signals will be present when a command-guided missile is in flight. There may be subtle differences in the guidance signal as seen from the actual target of the missile—as opposed to sensors on another friendly platform (not the target) that can receive the guidance signal associated with a missile attacking the target.

Fusing

Fusing radars are very short range—a few times the burst radius of the weapon. They are independent of the other radars, and will be turned on at a distance that depends on the type of missile.

6.1.2   Track-While-Scan Threats

Track-while-scan (TWS) threats are different in that the search and acquisition processes are performed by the same radar, and simultaneously with the tracking of the threat. This allows the radar to track multiple targets while searching for more.

The target being tracked will not detect any change in the tracking radar when it has been selected as a target.

6.1.3   Antiship Missiles

Antiship missiles are normally fired from a great distance. They can be launched from ships on targeting information from some remote observation asset (aircraft, satellite, etc.). They can also be launched from aircraft that acquire targets with their own search radars. Figure 6.2 shows the phases of a typical antiship-missile attack.

Once launched, the missile guides itself inertially to the target location designated before launch. The missile skims the surface during its approach and does not radiate. Common missiles are either slightly subsonic or supersonic at about mach 2.5. From a surface-skimming altitude, the ship breaks horizon at about 10 km.

When the ship is expected to break horizon, the antiship missile turns on its radar and tries to acquire the ship. When it detects the ship, the radar locks on and guides the missile toward the center of the ship.

During the terminal phase of the attack, the missile either dives to strike the ship at the waterline or climbs for an almost vertical attack straight down through the deck.

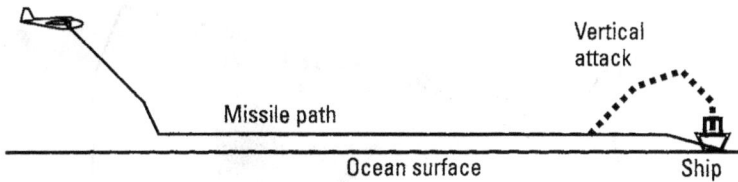

Figure 6.2 Antiship missile flight path.

6.1.4 Air-to-Air Threats

Tactical aircraft are normally vectored into attack position by combat controllers based on information from airborne or ground-based radars, as shown in Figure 6.3 Long-range ground radars are often called *early warning/ground-control intercept* (EW/GCI) radars for this reason. The airborne radars used to detect and establish tracks upon target aircraft are very similar. Neither the airborne nor ground-based acquisition radars will change modes when they detect targets, since they handle multiple airborne targets and continue to search for more targets.

When a fighter with an AI radar is being vectored toward a target, a digital control signal may be present. This is particularly true in situations in which there are only narrow corridors through which aircraft can fly to avoid shoot-down by ground-based antiair defenses.

Once it is within range, the fighter will turn on its fire control radar, which has several modes. Usually, the fire control radar will cover a large angular area to acquire its target. Once locked to a target, it will change to a tracking mode in which it covers a much narrower area and uses a different modulation.

The fire-control radar transmits over a limited angular segment from the nose of the aircraft, as shown in Figure 6.4. This signal can be received at full power only when the target aircraft is within the lethal zone, but may be

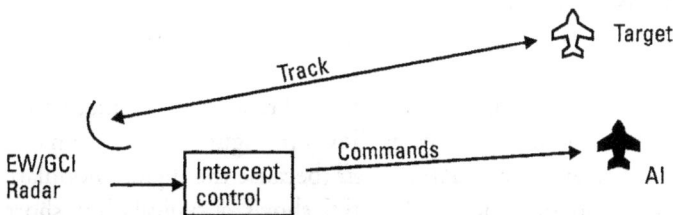

Figure 6.3 Radar-controlled aerial intercept.

Figure 6.4 Lethal zones for AI radars.

received at side-lobe level from other angles. There are generally two lethal zones: one would allow a launch to take place and the second, smaller area would typically not allow the target to escape. In general, a launch can be expected to occur when a target comes into the edge of the no escape zone.

6.2 Modulations

There are two basic types of signal modulations of concern to EW systems and operations. One is pulsed and the second is continuous. Pulse signals are associated with radars, while continuous signals can be associated with either radar or communication.

6.2.1 Pulsed Signals

Pulsed signals are 100% amplitude modulated, and are characterized by RF frequency, pulse width, duty factor, pulse pattern, and modulation on the pulses.

In general, the longer the range, the lower the frequency, the longer the pulse width, and the longer the pulse interval. Search radars typically have pulse widths of several microseconds. The long pulses allow the radar to get more energy onto the target—which enhances the detection range. However, this degrades their range resolution. Frequency or binary modulation is often added to reduce the range resolution cell. Frequency modulation on pulse is called *chirp*. Its implementation is shown in Figure 6.5. The binary modulation is often referred to as *Barker code* (because the digital modulation often forms that kind of code). Figure 6.6 shows a typical, but short, binary sequence on a pulse. The pulse repetition interval of search radars can be expected to be several milliseconds.

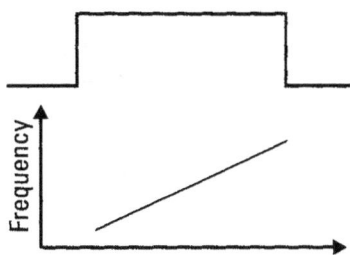

Figure 6.5 Linear FM on pulse.

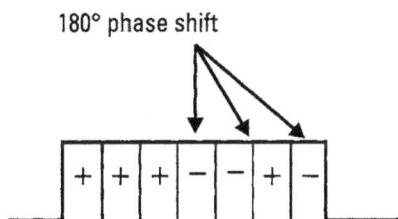

Figure 6.6 Binary code on pulse.

Tracking radars have much shorter pulse widths and higher pulse repetition frequencies. The pulse widths are from a very few microseconds down to sub-microsecond. The pulse repetition intervals are from a very few milliseconds to submilliseconds. Except for high PRF pulse Doppler radars, the duty cycle will be very low.

Airborne fire control radars often have a pulse Doppler mode in which the pulse repetition frequency may be about 300,000 pulses per second, and the duty cycle can be of the order of 30%.

Because receivers limit the bandwidth of signals they receive, the pulses are seen to be rounded on their leading and trailing edges. In general, transmitters are wider in bandwidth than the associated receivers, so transmitted pulses are usually considered to be nice square pulses.

6.2.2 Continuous Signals

Continuous Radar Signals

Tracking radars can have continuous-wave (CW) modulation. If it is truly a CW signal, there is no modulation present. This means that the radar cannot measure range to the target, but only the relative velocity of the radar and the target (because of Doppler shift). Of course, it can still determine the angular location of the target because of its narrow antenna beam.

If the radar signal is frequency modulated, it will be able to measure both range and relative velocity. The FM on CW radars will normally be a linear ramp. As shown in Figure 6.7, the frequency difference between the transmitted and reflected signals indicates the range to the target. However, the frequency difference will also be affected by the relative velocity, so there may be a modulation like that shown in Figure 6.8. The Doppler shift is measured during the flat part of the curve and the frequency difference measured during the ramp is corrected for Doppler shift.

Communication Signals

Communication signals have continuous modulation carrying continually changing information. These signals are typically AM, FM, or phase modulation (PM). They can carry either analog information or digital information.

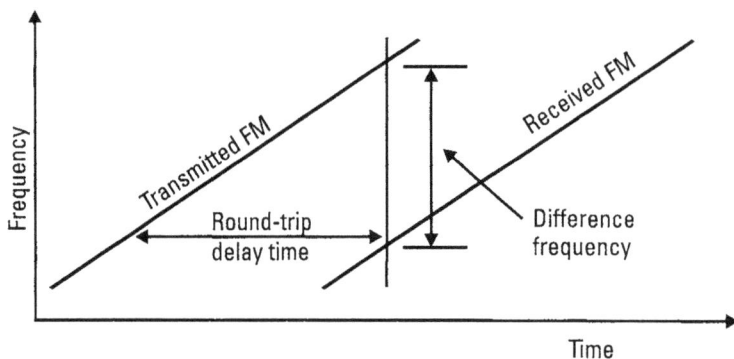

Figure 6.7 FM on CW radar for range measurement.

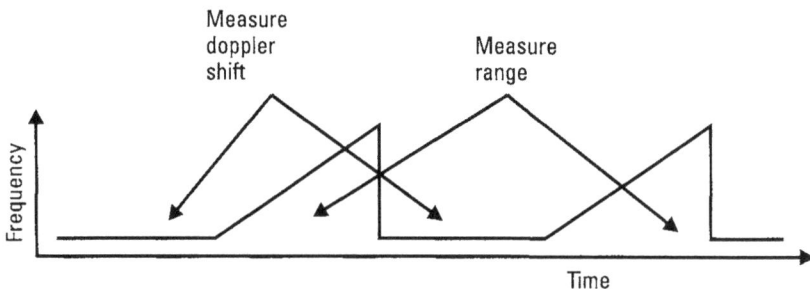

Figure 6.8 Ramp modulation allowing both Doppler and range measurements.

In the case of AM, the percentage of modulation must be specified. Modulation of 30% to 50% is typical. The transmission bandwidth of AM signals is twice the bandwidth of the modulating information. The transmission bandwidth of FM signals depends on the modulation index: transmission bandwidth equals twice the modulation index times the bandwidth of the information input to the transmitter. Digital signals can be modulated onto a carrier as AM, FM, or PM. If phase modulation is used, it is normally either binary phase shift keying (BPSK) or quadrature phase shift keying (QPSK). For BPSK, the transmission bandwidth is normally about equal to twice the data bit rate. For QPSK, it is about equal to the bit rate.

The modulation for a communications threat signal can be found from intelligence data. An unclassified example is the normal push to talk FM tactical radio. It has an information input of about 4 kHz and a transmission bandwidth that fits in a 25-kHz channel with appropriate guard bands (i.e., a modulation index of about 3).

6.3 Antenna Characteristics

6.3.1 Communication Threat Antennas

The antennas for most communication antennas transmit 360 in azimuth, and are received near the horizontal plane, so it is sufficient to just consider the ERP as the transmitter power increased by the gain. When received from straight overhead, the communication antenna will typically have a pattern null that may be as much as 20-dB deep. While this reduces the ERP considerably, it is not an important characteristic. A receiver directly overhead will not usually remain there very long (aircraft pass over quickly) and the receiver will be very close to the transmitter (unless it is in a satellite), so it will receive plenty of signal to do its job. The challenge usually comes because the receiver, particularly a noncooperative EW receiver, is far away near the horizon.

When a VHF or UHF communication antenna has directivity, it is normally a log-periodic or similar antenna with a pattern as described in Section 5.1.3. A directional communication antenna typically does not rotate during operation. It is set to optimally carry information from one fixed position to another.

For microwave links, the antennas are narrow-beam horns or dishes. They have gain patterns like radar antennas and are discussed in Section 5.1.3. They are different from radar antennas in that they do not change their orientation during operation. If you are in the beam, you get lots of signal and if you are out of it you receive only side-lobe level signals.

6.3.2 Radar Antennas

Radar antennas are more interesting; not only do they have narrow beams, but the beams move in ways that tell us a lot about the type of radar and its operating mode. The movement of the antenna is called the *antenna scan*. The real interest is how the threat antenna looks to an EW receiver as it is moved.

Threat Antenna Patterns As Seen by a Receiver

As discussed in Section 5.1, the gain pattern of an antenna is a plot of its gain as a function of the angle from its bore sight to the observation point. For a transmitting antenna, this causes an angular variation in its effective radiated power.

If the transmitting antenna is scanning, a time-varying signal level will be observed by a receiver in a fixed location, as shown in Figure 6.9. The main beam of the transmit antenna causes a peak signal to be received as it passes through the receiver location.

Each of the transmitter antenna side lobes causes a smaller peak signal as it passes the receiver location. Assuming that the transmit antenna is

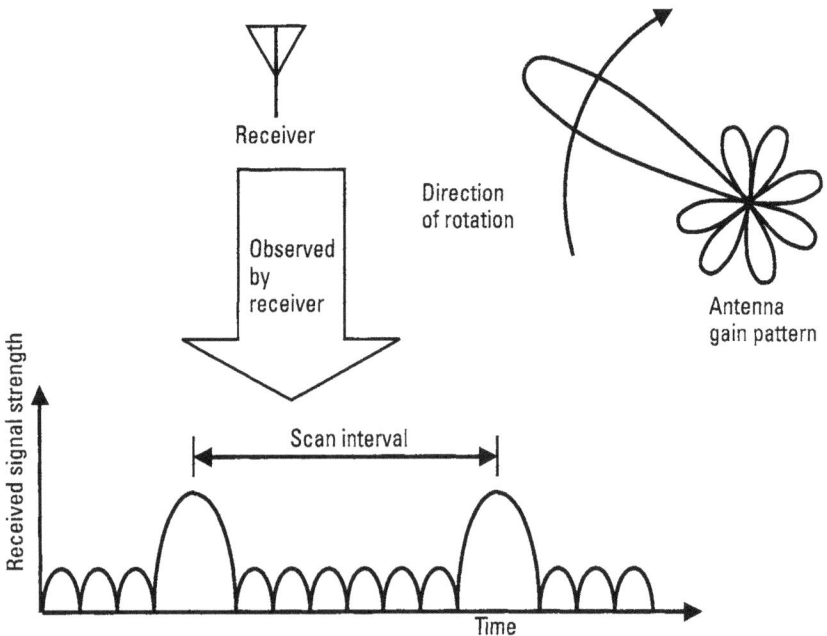

Figure 6.9 Antenna pattern as observed by a receiver.

rotating at a constant rate, each of these peak signals will be received at a fixed interval. As shown in the figure, the scan interval is the time between receipts of the main lobe peak by the receiver (i.e., a 5-second interval means the antenna is rotating once per 5 seconds).

As shown in Figure 6.10, the time between the 3-dB points on the received main beam signal strength curve is related to the 3-dB beam width of the transmitting antenna. The relationship is

Beam width = (beam duration / scan period) × 360 degrees

There are a number of types of radar antenna scans, depending on the type of radar, its mission, and its operating mode. Analysis of the antenna pattern is one of the tools used by EW receivers to identify enemy radars. Also, the pattern will determine the amount and periodicity of the time during which a receiver can receive the threat signal. Thus, it is important for a simulator to accurately represent the received antenna patterns of simulated threats.

In the balance of this section, we will discuss a number of types of antenna scans. For each, we will consider what the threat antenna is doing (i.e., its angular movement history), the general mission of a radar using this scan pattern, and what it looks like to a receiver in a fixed location.

6.3.3 Circular Scan

As shown in Figure 6.11, the circularly scanned antenna rotates in a full circle. This type of scan is associated with search radars. It typically monitors a large area to detect targets, which are handed off to tracking radars if appropriate.

The received pattern is characterized by even time intervals between observations of the main lobe, as shown in Figure 6.12.

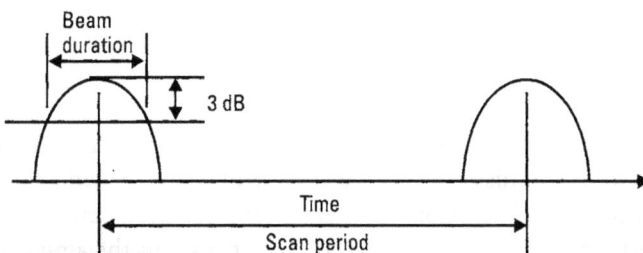

Figure 6.10 Beam duration and scan period of transmit antenna as observed by a receiver.

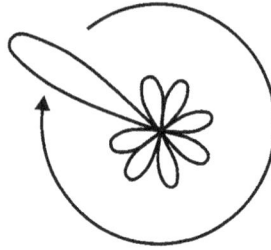

Figure 6.11 Circular scan.

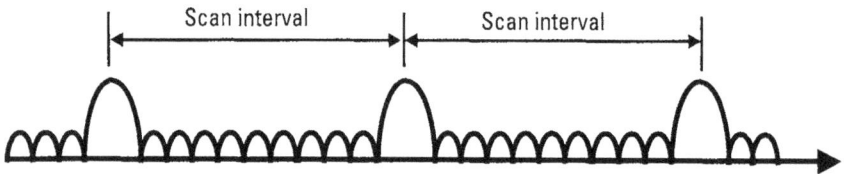

Figure 6.12 Received power versus time for circularly scanned antenna.

6.3.4 Sector Scan

The sector scan differs from the circular scan in that the antenna moves back and forth across a segment of angle. This scan concentrates its attention in a smaller angular segment, increasing the probability that it will intercept a short duration signal. It is used during the acquisition phase by antiship missile radars. The time interval between main lobes has two values, except in the case in which the receiver is at the center of the scan segment. Figure 6.13 shows the physical sector scan and Figure 6.14 shows the time power history as this type of scan is observed from a fixed receiving position.

6.3.5 Helical Scan

The helical scan covers 360 degrees of azimuth and changes its elevation from scan to scan. This scan covers a volume of space, providing both elevation and azimuth information—as well as range—about detected objects. It is observed with constant main lobe time intervals, but the amplitude of the main lobe decreases as the threat antenna elevation moves away from the elevation of the receiver location. Figure 6.15 shows the physical scan, and Figure 6.16 shows the observed power history.

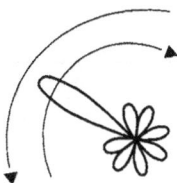

Figure 6.13 Sector scan.

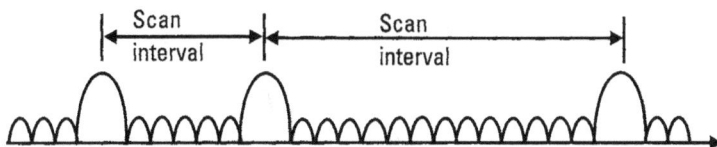

Figure 6.14 Sector-scan observed power history.

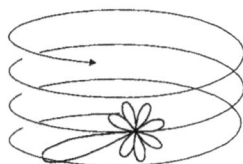

Figure 6.15 Helical scan.

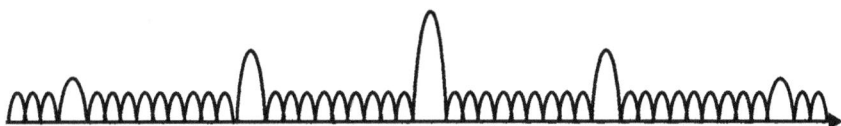

Figure 6.16 Helical-scan observed power history.

6.3.6 Raster Scan

The raster scan covers an angular area in parallel lines. This scan covers an angular volume, and is often used as the acquisition scan for airborne fire-control radars. It is observed as a sector scan, but with the amplitudes of the main lobe intercepts reduced as the threat antenna covers raster lines that do not pass through the receiver's location. Figure 6.17 shows the physical scan and Figure 6.18 shows the observed power history.

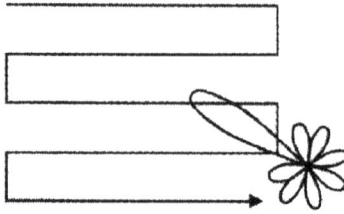

Figure 6.17 Raster scan.

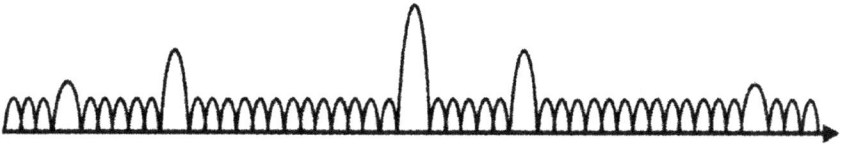

Figure 6.18 Raster-scan observed power history.

6.3.7 Conical Scan

The conical scan is observed as a sinusoidally varying waveform with a maximum when the antenna beam comes closest to the receiver's location and a minimum when it is pointed maximally away from the receiver. It is an excellent example of a target-tracking scan as the receiver location (T in the figure) moves toward the center of the cone formed by the scanning antenna, the amplitude of the sine wave reduces. When the receiver is centered in the cone, there is no variation in the signal amplitude, since the antenna remains equally offset from the receiver.

Figure 6.19 shows the conical scan with three target positions (A, B, and C) moving from the outside to the center. Figure 6.20 shows the corresponding power history observed by a receiver located on the target.

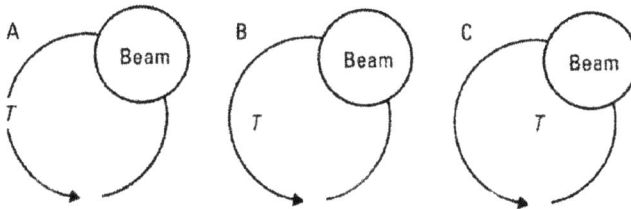

Figure 6.19 Conical scan as target moves toward center.

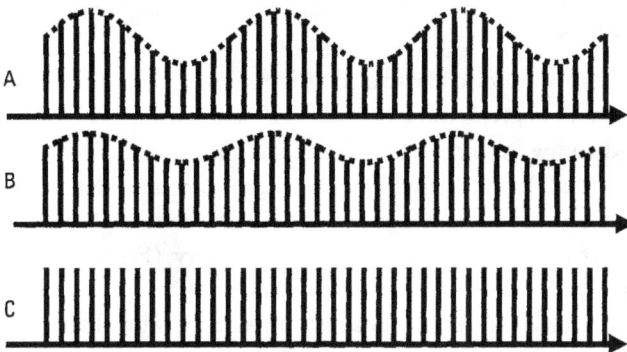

Figure 6.20 Conical-scan power history observed by target in three positions.

6.3.8 Spiral Scan

The spiral scan is like a conical scan, except that the angle of the cone increases or decreases. The observed pattern looks like a conical scan for the rotation, which passes through the receiver's location. The antenna gain diminishes in amplitude as the spiral path moves away from the receiver location, but still retains a generally sinusoidal shape. The irregularity of this pattern comes from the time history of the angle between the antenna beam and the receiver location. Figure 6.21 shows the physical scan and Figure 6.22 shows the observed power history.

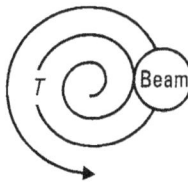

Figure 6.21 Spiral scan.

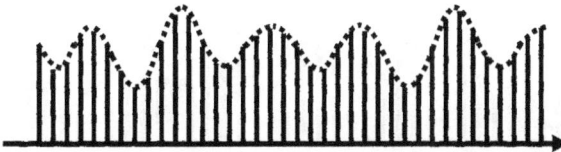

Figure 6.22 Spiral-scan power history observed from target position.

6.3.9 Palmer Scan

The Palmer scan is a circular scan that is moved linearly. The name comes from the old Palmer method of teaching handwriting, in which students started by drawing looping circles like those shown in the scan motion diagram (Figure 6.23). Its advantage is that it provides high-density coverage of an area with a small vertical angle and a larger horizontal angular segment. The observed threat antenna gain pattern history (Figure 6.24) is a very strange waveform. If the receiver were right in the middle of one of the circles, the amplitude would be constant for that rotation. In the figure, it is assumed that the receiver is close to the center, but not exactly centered. Therefore, the third cycle shown is a low-amplitude sine wave. As the cone moves away from the receiver location, the sine wave becomes full size, but the amplitude of the signal diminishes.

6.3.10 Palmer Raster Scan

If the conical scan is moved in a raster pattern, the received threat gain history will look like the Palmer scan for the line of the raster that moves through the receiver location. Otherwise, the pattern becomes almost sinusoidal, with diminishing amplitude as the raster lines move farther from the angle of the receiver location. Figure 6.25 shows the physical movement of the antenna and Figure 6.26 shows the observed power history.

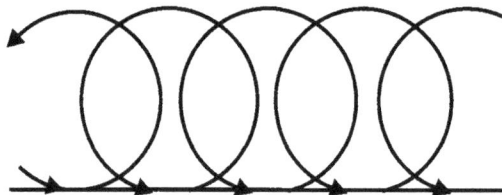

Figure 6.23 Palmer scan.

Figure 6.24 Palmer-scan observed power history.

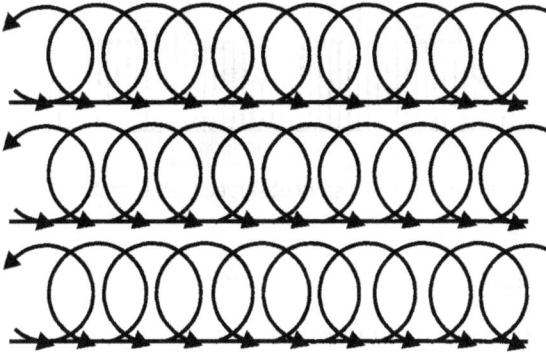

Figure 6.25 Palmer raster scan.

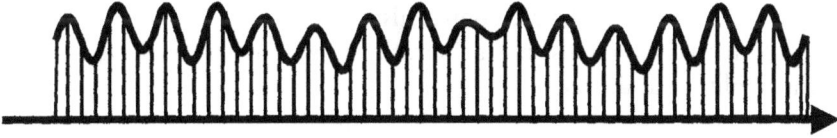

Figure 6.26 Palmer-raster-scan observed power history.

6.3.11 Lobe Switching

The antenna snaps between four pointing angles forming a square (in this case) to provide the information required to track its target. Like the other patterns, the received threat antenna gain history is a function of the angle between the threat antenna and the receiver's location. Figure 6.27 shows the four antenna beam lobe positions with the target in one position. Figure 6.28 shows the observed power history at the target in the position shown.

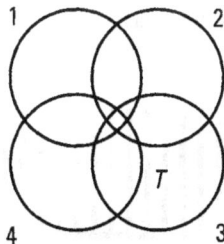

Figure 6.27 Lobe switching.

Figure 6.28 Lobe-switching power history observed from target.

6.3.12 Lobe on Receive Only

In this case, the threat radar tracks the target (the receiver location) and keeps its transmitting antenna pointed at the target if it has a directional antenna. The receiving antenna has lobe switching to provide tracking information. The receiver sees a constant signal level, since the transmit antenna is always pointed at it. Figure 6.29 shows the separate transmit and receive antennas. Figure 6.30 shows the observed power history.

Note that monopulse radars in tracking mode are also observed as constant amplitude at the target.

6.3.13 Phased Array

This assumes that the phased array is electronically steered, so it can randomly move from any pointing angle to any other pointing angle instantly

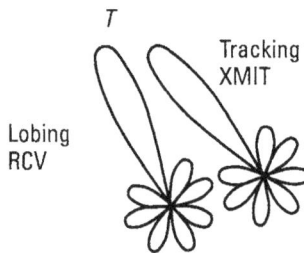

Figure 6.29 Lobe on receive-only antennas.

Figure 6.30 Lobe on receive-only observed power history.

(see Figure 6.31). Thus, there will be no logical amplitude history observed by the receiver. A fixed-beam phased array that is moved mechanically would, of course, look much like the appropriate one of the above-described scan patterns. The received gain depends on the angle between the instantaneous-threat-antenna pointing angle and the receiver location. Figure 6.32 shows the observed power history (i.e., completely random).

6.3.14 Electronic-Elevation Scan with Mechanical-Azimuth Scan

In this case, the threat antenna is assumed to have a circular scan with the elevation arbitrarily moved by a vertical phased array. Thus, there is a constant time interval between main lobes, but their amplitude can vary without any logical sequence. The azimuth scan can also be a sector scan, or it can be commanded to fixed azimuths. Figure 6.33 shows the physical movement of the scan and Figure 6.34 shows its observed power history.

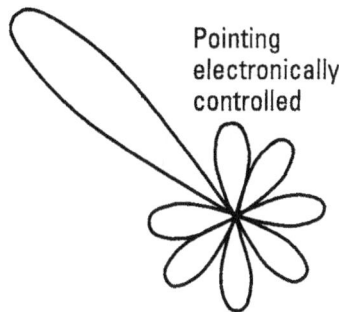

Pointing
electronically
controlled

Figure 6.31 Phased-array beam.

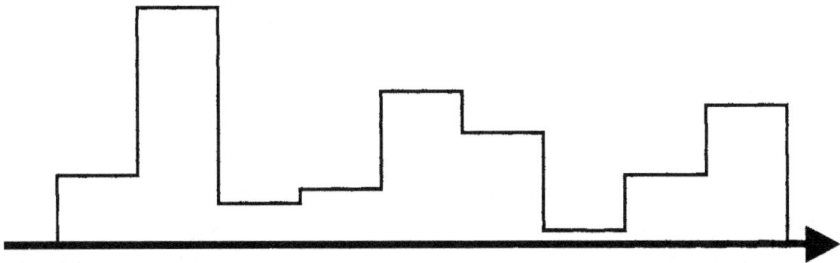

Figure 6.32 Random observed power history of electronically controlled phased array.

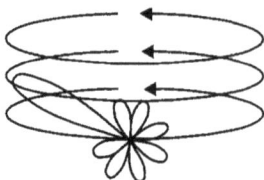

Figure 6.33 Electronic-elevation mechanical-azimuth phased array.

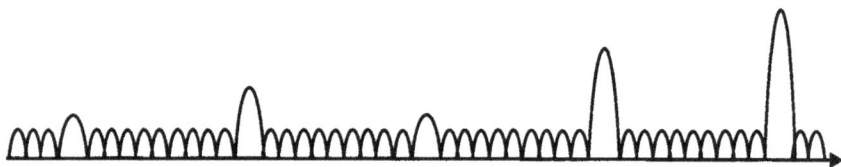

Figure 6.34 Observed power history of electronic-elevation mechanical-azimuth phased-array radar.

6.4   Signals Leaving Transmitter Site

The signals leaving the transmitter site have the applied modulation, and have an effective radiated power (ERP) that is the sum (in dB) of the transmitter power and the antenna gain. The antenna gain pattern causes signals to be radiated in all directions. The ERP is, of course, a function of the angle from the antenna bore sight.

The modulation is not changed by the off-bore-sight antenna gain reduction. It is only the radiated RF signal level that is reduced.

6.5   Signals Arriving at Receiving Site

Signals leaving the transmission site in the direction of the receiver have an ERP that is the product of the transmitter power and the antenna gain in the direction of the receiver. Taking the peak ERP at the transmit antenna bore-sight as the standard, the effective ERP of a scanning threat is reduced by appropriate attenuation versus time to reproduce the antenna scans shown in Section 6.3. For nonscanning threats, the ERP remains constant at the transmitter power plus the antenna gain in the direction of the receiver.

Figure 6.35 shows a typical scanning radar signal as it looks leaving the transmitting site. Note that the scanning pattern causes a time-varying reduction of the transmitted pulses.

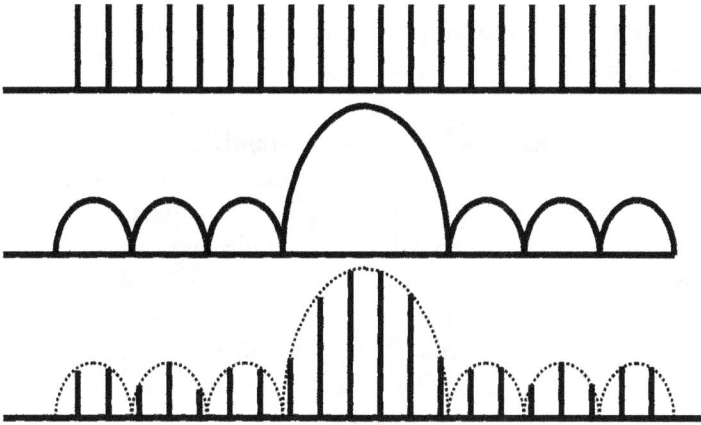

Figure 6.35 Scanning pulse radar signal leaving transmitter site in direction of receiver.

Once the signal leaves the transmit site, it is reduced by the link losses described in Chapter 5—if there is line of sight between the transmitter and the receiver. If the signal path is at any time interrupted by terrain, it is common practice to assume that the signal strength drops to zero. In Figure 6.36, receiver R2 can receive the signal, but receiver R1 does not have line of sight to the transmitter, and thus will receive no signal.

Another consideration is the horizon. Signals at VHF frequencies and above are normally assumed to be limited to the radio horizon, which is determined by the following formula:

$$D_{max} = 4.123\left(h_T^{\frac{1}{2}} + h_R^{\frac{1}{2}} \right)$$

where

D_{max} = the maximum RF line-of-sight distance from transmitter to receiver (or radar to target) (km);

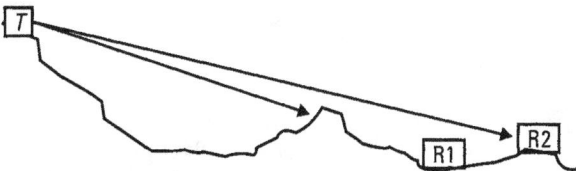

Figure 6.36 Terrain-masking of signals to receiver.

h_T = height of transmitter above sea level (m);

h_R = height of receiver above sea level (m).

This formula uses the common four-thirds Earth assumption with smooth Earth surface at sea level. The formula works reasonably well if the transmitter and receiver heights are above local level terrain. The equation also works with the two heights being a radar and a target.

The frequency of the signal arriving at the receiving site is altered by a Doppler shift if the transmitter or the receiver or both are moving. The change in frequency is proportional to the rate of change of distance between the transmitter and receiver. This is determined as shown in Figure 6.37. The angles θ_T and θ_R are the true spherical angles between the transmitter and receiver velocity vectors and the line-of-sight path between the two. This same concept applies to a radar-tracking a target, except that there is a factor of two on the right side of the equation to allow for two-way transmission of the radar signal.

It should be noted that the PRF and PW of pulsed signals are also modified by the Doppler shift.

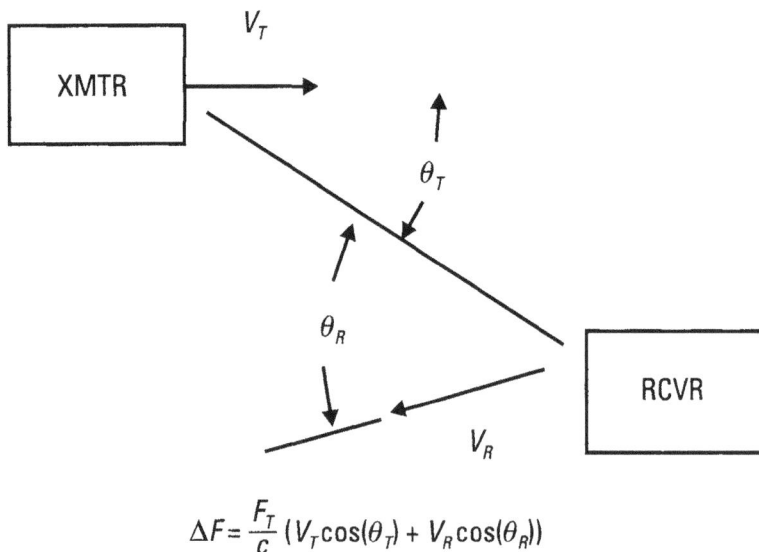

$$\Delta F = \frac{F_T}{c} \left(V_T \cos(\theta_T) + V_R \cos(\theta_R) \right)$$

Figure 6.37 Doppler effect for arbitrary velocity vectors.

7

Engagement Modeling

In the modeling of an EW engagement, it is necessary to consider the following elements:

- The characteristics of the area in which the engagement takes place (the gaming area);
- The characteristics of the players in the engagement;
- The way players move and react to each other during the engagement;
- The point of view from which the engagement is to be observed;
- The fidelity to which the engagement must be controlled;
- The electronic interactions between the players in the engagement;
- When to start and stop the engagement.

7.1 Gaming Area

The gaming area is the space in which the players engage each other. Note that the gaming area can have many dimensions: location, frequency, and so on. The steps in construction of the model are as follows:

1. *Design the gaming area.* How much area does the action cover (including the elevation of the highest player)? From how far away can a player affect another player? What other dimensions must be

considered in the model? (Usually time is included, and sometimes frequency or some other dimension must be considered.) What coordinate system is most convenient for the simulation to be run? Often a Cartesian coordinate system can be used (x and y along a flat zero altitude surface and z for elevation), with zero being at one corner of the gaming area. Sometimes it is convenient to use polar or Cartesian coordinates centered on one player.

2. *Add terrain elevations to the gaming area if appropriate.* Typically, a sea engagement is assumed to take place over a flat, sea-level surface. A high-altitude air-to-air engagement can usually ignore surface features. Ground-to-ground or air-to-ground engagements are usually very interactive with terrain.

The EW gaming area as shown in Figure 7.1 is the space in which the simulated action takes place. It must be large enough to include all of the locations that will be occupied by the players in the simulation (threats and EW protected platforms). It usually has an orthogonal coordinate system that allows each player to be located in x, y, and z. This makes it possible to calculate relative angles and distances between the players during the simulated engagement.

The gaming area does not need to be three-dimensional. For example, in an engagement between a ship and an antiship missile, we know the elevation of the ship—it is at sea level. The tracking action is in two dimensions because the missile knows the elevation of the ship. Therefore, the gaming area can be two-dimensional.

The coordinates and origin of the gaming area need not be as shown. The origin can be a defended asset, with all attacks and countermeasures being in locations relative to the defended asset. The coordinates could be polar from that origin, in order to simplify the math.

7.2 Players

The players in the EW model include the following:

- All friendly assets, for example an aircraft penetrating enemy airspace, a ship that is under attack, supporting aircraft, ships, or ground assets (all of which emit signals);

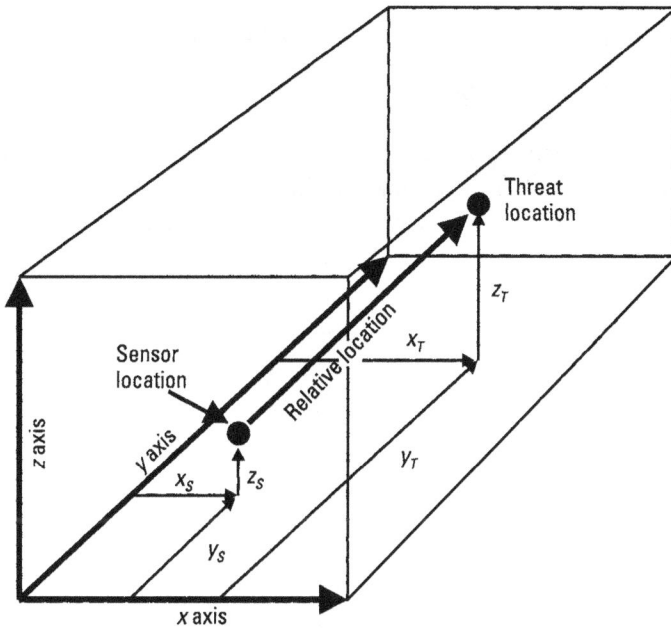

Figure 7.1 Typical EW model gaming area.

- All threat-platform and fixed-site facilities, which will be emitting signals or threatening friendly assets, for example ships (with radar and communications), aircraft (with radar and communications), SAM sites, AAA sites, tactical vehicles with communication transmitters, missiles (with or without radars).

Each player must be characterized. What are its qualities? What countermeasures and counter-countermeasures are used? What specific actions by other players cause changes in those qualities? How does it move through the gaming area? Each of these qualities must be described numerically, and the movements must be described in terms of equations, with the specific actions (or results of those actions) as elements of the equations.

For a model to yield the correct information, it is necessary that the qualities of the players be well described. This typically involves some rules of reaction. For example, an antiship missile does the following:

- Turns on its radar at the horizon (10 km from the ship) at an angle that is input to the model;

- Has an initial velocity vector (speed and azimuth) that is input to the model;
- Maintains constant speed throughout the engagement;
- Always steers toward the point that represents the weighted sum of all RCSs within the resolution cell of its radar, and keeps its resolution cell centered at the location of that weighted sum;
- Has an onboard radar with a fan beam antenna that has a horizontal beam width of 5 degrees and a peak gain of 23 dB. Its ERP at the antenna bore sight is +100 dBm;
- Cannot distinguish between skin returns from the target, returns from decoys, and returns from chaff;
- Updates its steering information once per second until it is within one second of the ship. Then it updates its position 10 times per second.

Another example that would apply to an aircraft penetrating enemy airspace is the action of an enemy fighter. The fighter does the following:

- Attacks directly from the rear and 100m below the friendly aircraft;
- Closes at 500 m/s;
- Can turn at 5 gs;
- Has a "J-Bird" radar;
- Turns on its radar in acquisition mode 10 km from the friendly aircraft;
- Acquires and starts tracking the largest RCS it sees (greater than 10 m^2) at 3 km;
- Tracks the weighted vector sum of all RCSs within its resolution cell and keeps the resolution cell centered on the point at which the weighted sum is located.

A fixed-site threat would, of course remain in its location throughout the engagement. However, it would change its operating modes as a function of distance to the protected asset, as described in Chapter 5. The threat-emitter parameters for each operating mode are assigned as is also described in Chapter 5.

For a ground mobile threat, its movement would normally be described in terms of a route (e.g., along a particular road or a straight path at some vector) and a travel speed. Its emitter characteristics would be described just as for a fixed threat.

An EW model usually focuses on a friendly asset that is being protected with EW assets. It can either follow rules through the engagement being modeled or it can be controlled by an operator. There can also be operators on the threat side. A simulation with real-time operator input is called a *man-in-the-loop* simulation. The model that supports that simulation is then impacted by those inputs. The operator commands must be converted into platform locations and orientations, system operating modes, antenna orientations, and so on. In any case, such things as the maximum speed, elevation, and turning rate must be defined for the friendly platform.

The EW assets must also be defined. What types of sensors are available? What types of defensive measures (jamming, chaff, etc.) are operative? What is the sensitivity of each sensor? What type of antenna scanning does it employ? What kind of threat density and distribution can it handle?

An important quality of the protected asset is its RCS. This can be in the form of a lookup table that has measured data for RCS versus aspect angle (and perhaps versus frequency) from an actual asset. It can alternately be in the form of an equation of RCS versus aspect angle, as shown in Figure 7.2 for a protected ship.

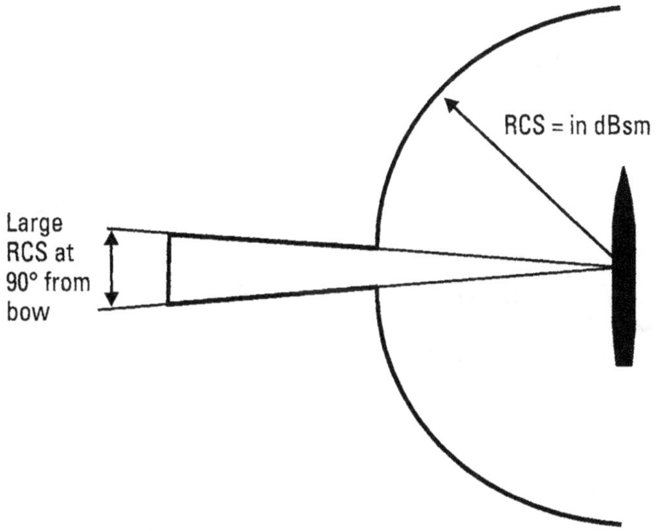

Figure 7.2 Mathematically described RCS versus aspect angle for a ship.

7.3   Location and Movement of Players

One of the tasks in modeling is to establish a threat laydown. This is literally the locations of threat emitters at the beginning of the engagement and the programmed changing of those locations as the engagement proceeds.

This will normally be a realistic picture of the way an enemy would use its emitting assets, as shown in Figure 7.3 (an admittedly simple example). It may even be a set of actual threat locations from reconnaissance data.

When the defended asset (e.g., the attacking aircraft) moves through the gaming area, it will encounter the threats in the laydown in a realistic way. The relative locations of the threats (distance and angle) and the timing of encountering the threats will be dictated by the characteristics of the threats and the sensors on the protected assets. The threat density and the instantaneous mix of threat types will follow naturally and realistically from the engagement geometry.

If the purpose of the engagement is to determine the relative value of various types of countermeasures, it may be sufficient to create a linear lay-down, as shown in Figure 7.4. In this case, the protected asset moves in a straight line past a variety of threat emitters that are placed in appropriate positions relative to that line. The mode change rules for each threat type are applied, and each launches its weapon(s) according to some rule. (For exam-ple, the weapon will launch at three-quarters of lethal range and will have a probability of kill of 50%.) Then, the effects of different types of counter-measures can be evaluated in terms of the probability that the asset will sur-vive the run through the threat laydown.

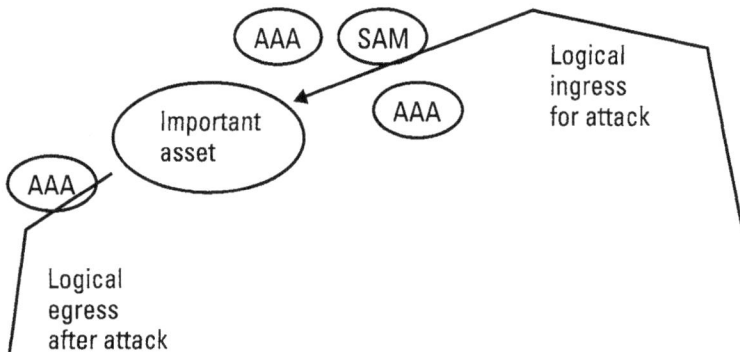

Figure 7.3 Threat laydown example.

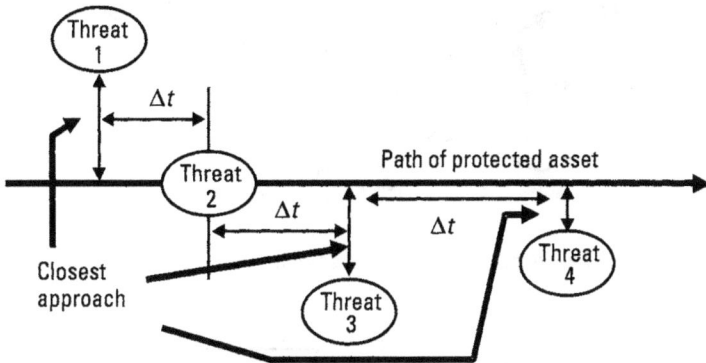

Figure 7.4 Threat laydown for comparison of EW asset benefits.

7.4 Point of View

The point of view of the model depends on what the model is intended to accomplish. If the purpose is to determine the outcome of an engagement, it is most common to use an all-knowing viewpoint. That is, to look at the action from above and know everything that is happening: exactly where every player is located, the mode of operation of every player, and so on. This is the viewpoint that must always be taken by the individual in setting up or running any model.

On the other hand, if the purpose of the model is to support a simulation, the point of view must be transitioned to the point of view of the sensor input that is to receive the simulated information. For example, as shown in Figure 7.5, a narrowband receiver is to receive signals from a number of threat emitters through a scanning antenna. In this case, the model of what the receiver will see depends on the instantaneous position of the antenna, and the tuning commands issued by an operator. The received signal power will be reduced by the off-axis antenna gain and it will only see the signal if it is tuned to the frequency of that signal.

Changing the point of view normally requires a change in the coordinates of the model. For example, if the threats and the sensors are laid out in orthogonal coordinates, and the object is to simulate inputs to a sensor (say a receiver), it will be necessary to determine the receiving antenna gain to every threat-emitter location. Since the antenna gain is known in spherical polar coordinates, the threat locations must be transformed to polar coordinates, with origin at the receiving antenna and the zero angle direction along the bore-sight axis of the antenna, as shown in Figure 7.6. Note that the antenna

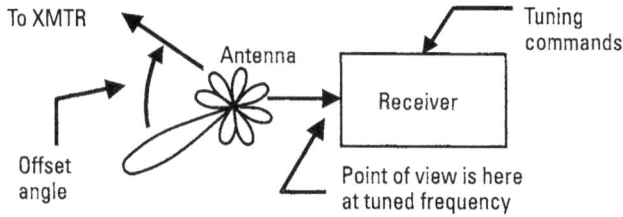

Figure 7.5 Point of view of narrowband receiver with scanning narrow-beam antenna.

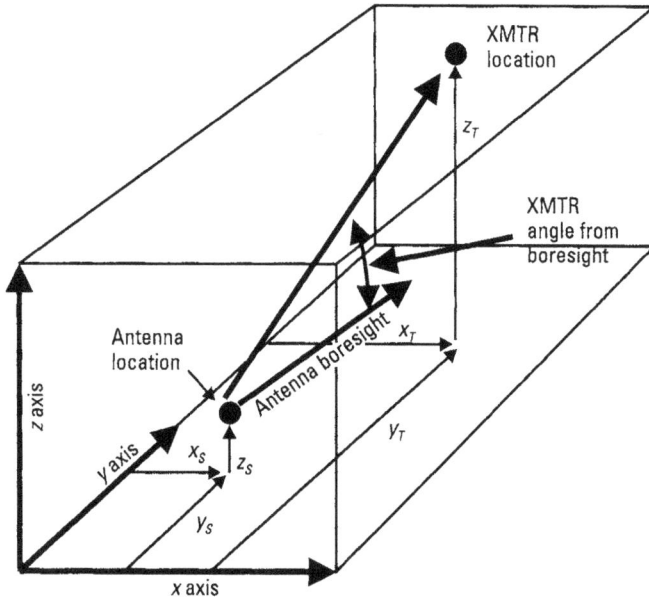

Figure 7.6 Coordinate transformation for receiver point of view.

bore-sight is parallel to the y axis (this is somewhat difficult to see in the drawing). In the new coordinate system, the transmitter is at range = sqrt($\Delta x^2 + \Delta y^2 + \Delta z^2$) and is at an elevation angle = arc sin [Δz/sqrt($\Delta x^2 + \Delta z^2$)] and a horizontal angle = arc sin [Δx/sqrt($\Delta x^2 + \Delta y^2$).

7.5   Engagement Fidelity

The fidelity level used in an engagement has a significant impact on the amount of computer power required to run the engagement. Too much fidelity is a waste of time and money; too little fidelity will reduce the validity of the results of the engagement modeling. The dimensions for which fidelity must be considered are as follows:

- The locations of all players;
- The time increments of calculation updates;
- The velocity vectors of all players (speed and direction);
- The RCS models of protected assets;
- The signal parameters of all emitters (frequency, ERP, antenna patterns, modulations);
- The orientation of antennas.

One fairly good way to set the appropriate fidelity level of the various aspects of the model is to start with the required accuracy of the results, then map the results resolution back to each model parameter resolution.

An example of an approach to this part of the analysis is to determine how closely you need to know the locations of the players, then divide the speed of the fastest-moving player by that location accuracy to determine the time increment at which calculations need to be made in the model.

There may be other requirements that dictate the accuracy of certain parameters. For example, the model for EW threat locations may need to be indexed to their image locations for the visual displays in a flight simulator. Typical examples of model resolution are as follows:

- Frequency to 1 MHz;
- Location to 1m; Speed to 1 m/s;
- Signal strength to 1 dBm;
- Modulation only to modulation type;
- RCS to 10%.

7.6 Electronic Interactions Between Players

Electronic interaction between two players typically involves a transmission from one player's location to a received signal at another player's location. A second typical interaction is a radar in the threat location receiving a return signal from a target at another location. The propagation of the signals is modeled using the formulas in Chapter 4 and the line-of-sight considerations discussed in Section 6.5.

Examples of the calculations involved in these interactions will be part of the two engagement examples at the end of this chapter.

7.7   Running the Engagement

Picking a Start Time

To run the engagement, first pick an appropriate starting point. Normally, this is when the first signal reaches the first sensor. This is practical in ground-to-ground or blue water naval engagements, since signals can be assumed to start when a threat is at the horizon. However, in some cases (for example, an aircraft entering hostile airspace) there will be many signals that can be seen from great range. In these cases, you would start the engagement at some time before the aircraft would be acquired by the first threat radar.

For the case of a system monitoring a signal environment, in developing an electronic order of battle (EOB), the starting point is somewhat arbitrary. A good example of this is the portion of a battlefield that is in the field of responsibility of a collection system like the army's GUARDRAIL. There are many tactical communication nets and radars, all of which have been operating before the collection system comes onto the scene. There is just no logical starting point. However, depending on the collection system, you might have a set of emitters and locations in system memory as part of the starting conditions.

Updating Engagement Conditions

Depending on the nature of the systems involved, the situation must be reviewed periodically and new calculations made. This update may or may not be counted in real time.

If the engagement model is to determine an outcome, the update time is time in the modeled situation. For example, if a missile is attacking a ship or aircraft, model time is time of flight of the missile, which may require either more or less time to calculate depending on the computer being used.

If the engagement model is driving a simulation, the update calculations must be often enough to avoid impacting the output resolution requirement. Do not confuse the recalculation interval with the output refresh rate. For example, in a graphical operator display, the operator may only be able to detect changes of two millimeters on the screen. If it takes 10 seconds for the tactical situation to change enough to cause that 2-mm movement on the screen, the recalculations need to be made only every 10 seconds—but the screen needs to be updated at least 24 times per second to avoid perceptible flicker. In this case, the screen updates would be made using the latest information from computer memory.

If the model is driving a simulator to test equipment, the measurements must typically be made in real time, since the equipment must be able to receive and process signals as they occur. The update increments are now driven by machine response time, which may be much quicker than human response time.

Ending the Engagement Model

For a model to determine outcomes, the logical end point is when you have the answer. The ship defense model is a good example. The object is normally to determine if the missile hits or misses the ship. If you are testing for distance of the missile from the middle of the ship, you will normally see a minimum at the time of closest approach. If the distance starts increasing, you can end the engagement. At that point, the missile has either hit or missed the ship.

For the example of the aircraft in hostile airspace, the problem ends when the aircraft leaves the engagement area. For the example of the system to develop EOB, you will probably just end after some arbitrary preset period of time (e.g., 5 minutes). If the model is driving a simulation, you quit when the test is done or the training lesson is completed.

7.8 Aircraft in Hostile Airspace

The purpose of this engagement scenario is to show the display on a radar-warning-receiver (RWR) vector scope as an aircraft flies an attack mission against a protected target. The output from this scenario could be used to generate a signal environment that would drive an emulation input to an actual receiver to test it under operational performance conditions.

Gaming Area

The gaming area includes a flat plane covering 100 × 200 km with two hills, and airspace up to 10 km above the flat plane, as shown in Figure 7.7. The coordinate system is orthogonal, with the origin at the lower left corner of the figure. The contour lines on the two hills denote 400m intervals above the flat plane; one hill is 1,000m high and the other is 1,700m high.

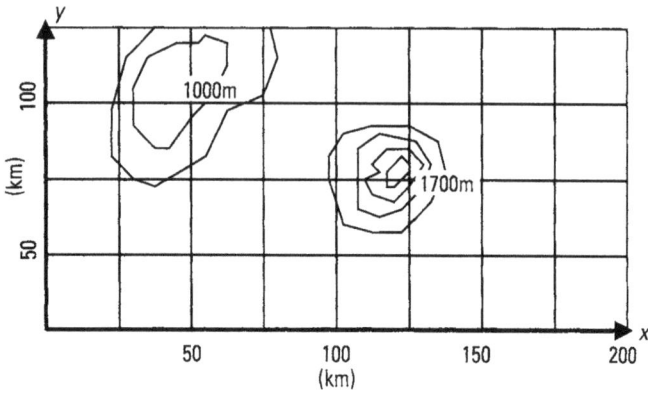

Figure 7.7 Example gaming area for aircraft in enemy airspace.

Players

The players, shown in the gaming area in Figure 7.8, are a friendly aircraft, an enemy fighter, three Gun Dish AAA sites, an SA-2 radar, an SA-3 radar, and a Flat Face early warning and target acquisition radar.

The attacking aircraft flies a constant 800-km/h. It passes through way-points in the gaming area at times T1 through T8. It follows an altitude contour (above the flat plane) as shown in Figure 7.9. It has an RWR that instantaneously covers 360 degrees of azimuth, and a sensitivity of –65 dBm. Its antenna gain is 0 dB. The RWR vector display is shown in Figure 7.10.

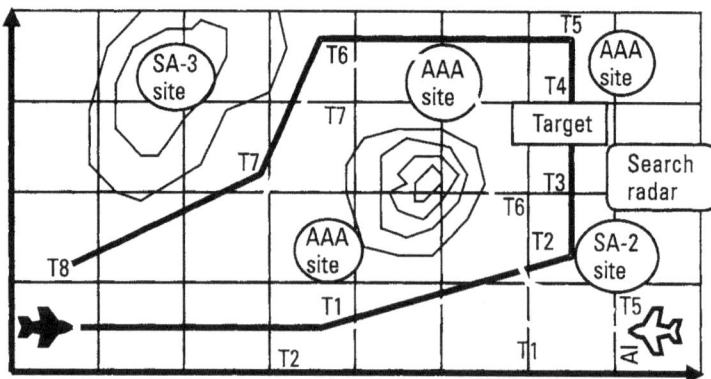

Figure 7.8 Gaming area with players.

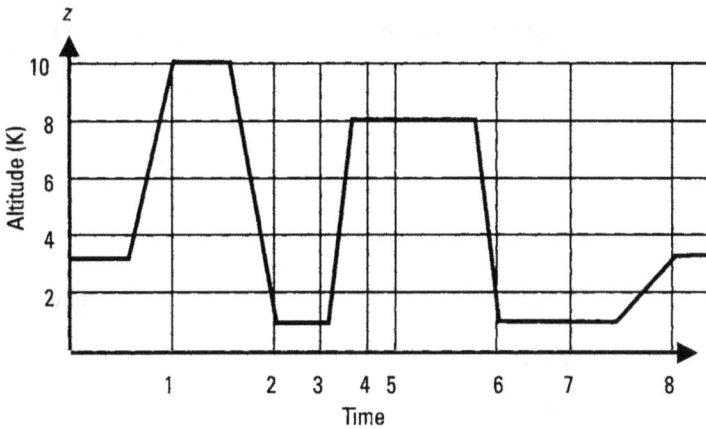

Figure 7.9 Elevation profile of aircraft.

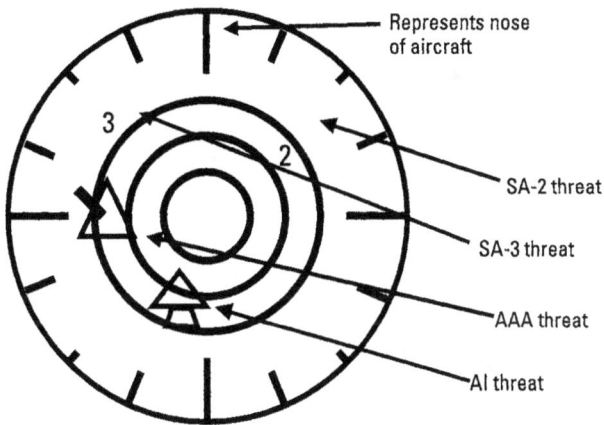

Figure 7.10 RWR vector scope.

The enemy fighter flies the two tracks shown in Figure 7.8. It is located at the points indicated by T1 to T7 at the time the attacking aircraft is at the corresponding way-points. The fighter has a Jaybird radar with the following parameters:

Frequency = 13 GHz;

ERP = 100 kW;

Antenna gain = 33 dB.

The Gun Dish AAA sites have radars with the following parameters:

Frequency = 15 GHz;
ERP = 100 kW;
Antenna gain = 43 dB;
Range = 8 km.

The SA-2 missile has a Fan Song radar with the following parameters:

Frequency = 5 GHz;
ERP = +125 dBm;
Antenna Gain = 34 dB;
Range = 100 km.

The SA-3 missile has a Low-Blow radar with the following parameters:

Frequency = 9 GHz;
ERP = +116 dBm;
Antenna gain = 32 dB.
Range = 50 km.

The Flat Face radar has the following parameters:

Frequency = 900 MHz;
ERP = +118 dBm;
Effective range = 225 km;
Antenna gain = 35 dB.

Point of View

The point of view for this engagement scenario is the pilot of the attacking aircraft, who views the vector scope of the RWR. The vector scope is shown in Figure 7.10. The symbols on the screen show the various threats and their momentary location relative to the aircraft.

Player Movement

The friendly aircraft enters the gaming area near the origin of the coordinate system at 3-km altitude. It climbs to 10 km at T1 to avoid the AAA site. It goes down to 800m at T2 and remains there until it passes T3, then it climbs to 8 km to attack the target. It goes back down to 800m at T6 and remains at that altitude until it is well past T7. Then it climbs back to 3 km to exit the gaming area.

The enemy fighter makes a head-on attack as the friendly aircraft passes T1. Then it makes a tail attack as the aircraft reaches T7.

Calculations

The calculations for time T4 are shown here as an example of how the process is performed. If the model were driving an actual simulator, these calculations would need to be repeated about once per second.

Threats up at T4:

AAA at 70 degrees at 13 km – out of range

SA2 at 170 degrees at 50 km – in range

Search at 270 degrees at 35 km – in range

Received Power from each threat (with receiving antenna gain assumed 0 dBi):

AAA $P_R = +80 - 32 - 20 \log(13) - 20 \log(15,000) = -58$ dBm

SA2 $P_R = +125 - 32 - 20 \log(50) - 20 \log(5,000) = -15$ dBm in the main beam. In the side lobes, it is 34 dB less = –49 dBm

Search $P_R = +118 - 32 - 20 \log(35) - 20 \log(900) = -4$ dBm in the main beam. In the side lobes, it is 35 dB less = –39 dBm

Results

The results of this engagement scenario are the screen displays on the RWR vector scope and the definition of the signals input to the RWR at each calculation update point. As an exercise, you can, if you like, make the calculations at each of the eight way-points shown in Figure 7.8 and record the results on the worksheet in Figure 7.11.

7.9 Ship Attacked by Antiship Missile

The purpose of this engagement scenario is to evaluate chaff as a defensive measure against an attack by an antiship missile. The scenario starts with a specific type of missile and a specific type of ship. The missile turns on its radar at the radio horizon and attempts to guide itself to the center of the

T1	T2	T3	T4
Threat ———	Threat ———	Threat ———	Threat ———
Range ———	Range ———	Range ———	Range ———
Angle ———	Angle ———	Angle ———	Angle ———
In range? __	In range? __	In range? __	In range? __
Threat ———	Threat ———	Threat ———	Threat ———
Range ———	Range ———	Range ———	Range ———
Angle ———	Angle ———	Angle ———	Angle ———
In range? __	In range? __	In range? __	In range? __
Threat ———	Threat ———	Threat ———	Threat ———
Range ———	Range ———	Range ———	Range ———
Angle ———	Angle ———	Angle ———	Angle ———
In range? __	In range? __	In range? __	In range? __

T5	T6	T7	T8
Threat ———	Threat ———	Threat ———	Threat ———
Range ———	Range ———	Range ———	Range ———
Angle ———	Angle ———	Angle ———	Angle ———
In range? __	In range? __	In range? __	In range? __
Threat ———	Threat ———	Threat ———	Threat ———
Range ———	Range ———	Range ———	Range ———
Angle ———	Angle ———	Angle ———	Angle ———
In range? __	In range? __	In range? __	In range? __
Threat ———	Threat ———	Threat ———	Threat ———
Range ———	Range ———	Range ———	Range ———
Angle ———	Angle ———	Angle ———	Angle ———
In range? __	In range? __	In range? __	In range? __

Figure 7.11 Aircraft in enemy airspace—worksheet.

ship. The ship maneuvers and uses chaff as a countermeasure. The scenario continues until the missile is past the ship. The distance from the ship to the missile is continuously calculated; if and when it becomes less than the size of the ship, the missile is understood to have hit the ship.

Players

The players in this scenario are the ship, the missile, and a chaff cloud. These are shown in Figure 7.12.

The ship has an RCS that is starboard/port symmetrical, but varies as shown in Figure 7.13 with the angle from the bow. The ship is proceeding due north at 50 km/h. Although the ship could maneuver during the engagement, a straight path is chosen.

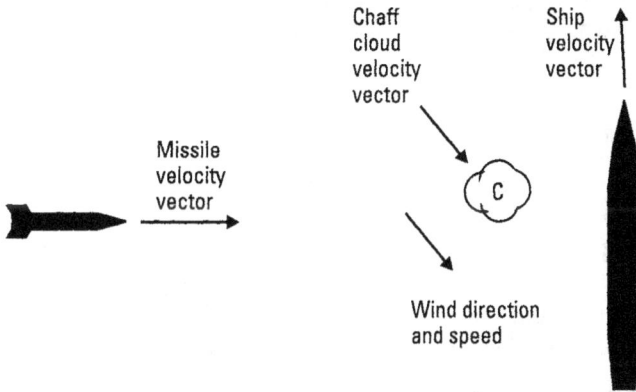

Figure 7.12 Players in ship-protection engagement.

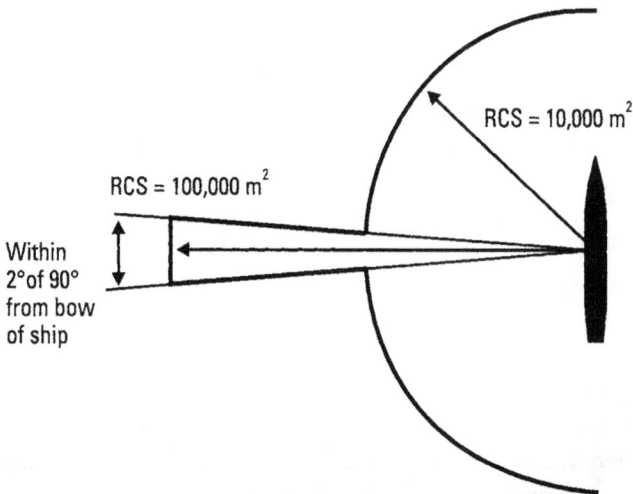

Figure 7.13 Model of ship's RCS.

The missile is a sea-skimming antiship missile that travels at a constant 1,000 km/h. It is actively guided with a 5-GHz radar with a fan beam antenna. The vertical beam-width is 30 degrees and the horizontal beam-width is 5 degrees. The radar ERP is 10 mW, and its pulse width is 1sec. The radar turns on 10 km from the ship. It guides the missile toward the center of the radar resolution cell.

The chaff cloud blooms to an effective RCS of 40,000 m². It drifts with the wind. Actually, it takes several seconds after the missile radar turns on for the chaff rocket to be launched and bloom, but for simplicity of explanation, we will place it in the ideal location one second after the beginning of the problem.

Gaming Area

In this scenario, the gaming area is two-dimensional, because it is clearly a two-dimensional engagement. (The missile knows the elevation of the ship.) The gaming area typically extends to the ship's radio horizon, a circle of about 10-km radius centered at or near the ship.

There is a wind across the gaming area. Both its direction and speed are inputs to the scenario. For this case, it is set at 10 km/h from the north/northwest (from azimuth 315 degrees).

The coordinate system is orthogonal (x and y), the origin is the center of the ship at the beginning of the problem. Note that there are other choices for the origin. For example, it could be maintained at the center of the ship.

Point of View

The point of view for this engagement scenario is from above. The observer knows everything that is going on. The locations of all of the players are recalculated once per second until the missile gets within 1 km of the ship, then the positions are recalculated 10 times per second.

The Engagement

All locations are given in (x, y) expressed in meters. The missile starts at (−10,000, 0) with its velocity vector azimuth at 90 degrees, the ship starts at (0.0), and the chaff cloud is deployed a few meters inside the downwind corner of the radar's resolution cell.

First, it is necessary to calculate the size and location of the resolution cell as shown in Figure 7.14. Its width is the distance from the missile to the ship (in meters) times 2(tan[BW/2]). Its depth is the speed of light (c) times (pulse width/2). At 0 seconds (the start of the problem), the cell is 873m wide and 150m deep, centered at (0,0). Therefore, the chaff cloud would

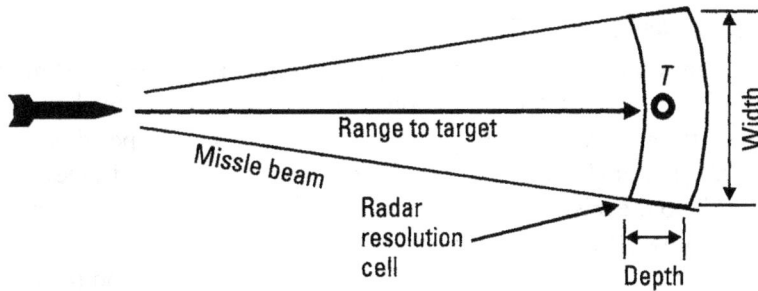

Figure 7.14 Radar resolution cell.

ideally be placed at (60,–420). The angle to the missile from the bow of the ship is 270 degrees, so the RCS is +50 dBsm.

Locations of the Players at Time = 1 Second

The ship has moved north at 30 km/h, which is (rounded) 8 m/s. At the end of the second, the ship is at (0,8). The missile has moved 278m on a vector of 90 degrees. Its position is (–9722,0). Its vector is 90 degrees – arc tan (8/9722) = 89.95 degrees, which is toward the center of the ship.

At one second, we need to calculate the radar resolution cell parameters to determine where to place the chaff cloud. The width of the cell is 9,722 (2 tan(2.5 degrees) = 849m. Its depth is c times (1μs / 2) = 150m. The chaff cloud is thus placed at (60,–400). Note that we want the chaff cloud just inside the downwind corner of the resolution cell, which is at (75,–417). (This is –425m + the 8m the ship has traveled north.)

The RCS in the direction of the missile is still 100,000 m^2, because the missile is 90.05 degrees from the bow of the ship (i.e., less than 2 degrees from the 90-degrees angle).

The missile's velocity vector is aimed at the center of its resolution cell, which is located between the center of the ship and the chaff cloud, proportional to the RCS ratio.

The RCS of the ship is 100,000 m^2 and that of the chaff cloud is 40,000 m^2, so the center of the resolution cell is 29% of the distance (40k/140k) from the ship to the chaff cloud. The center of the cell's x location is 60 × .29 = 17m. Its y location is 408 × .29 south of the ship = 110, so it is at –102. This puts the aiming point of the missile at (17,–102).

The missile's velocity vector is toward the center of its resolution cell, which orients the missile's velocity vector at 90 degrees + arc tan(102/9739) = 90.6 degrees. The ship and the missile are now separated by 9,722m.

Locations of the Players at Time = 2 Seconds

The ship moves another 8m north to (0,16). The missile moves 278m on a vector of 90.6 degrees, so its location is now (–9444,–3). The chaff cloud moves (with the wind) 3m along azimuth 135 degrees, so its position is now (62,–402). The angle from the bow to the missile is now 90.1 degrees, so the ship's RCS to the missile is still 100,000 m². The center of the radar's resolution cell is now (18,–105). The missile's new vector is still 90.6.

The ship and the missile are now separated by 9,444m, and the missile still sees the 90-degrees "ear" on the ship's RCS. However, the ship's motion will cause the aspect angle from the bow to increase as the distance closes. If the missile leaves that large RCS area, the center of the resolution cell will be 71% of the distance from the ship to the chaff cloud—which is drifting away from the ship by the combination of the ship's speed and the wind.

If the RCS to the missile drops while the chaff cloud is still in the resolution cell, the chaff cloud will capture the cell and the ship will soon be out of the cell. If that happens, the missile will guide directly toward the chaff cloud.

On the other hand, if the chaff cloud leaves the cell while the missile still sees the large ship RCS, it will home directly on the ship. Thus, it might have been a better strategy for the ship to turn away from the missile to increase the angle from the bow to the missile as quickly as possible.

Also note that the resolution cell is narrowing as the missile approaches the ship (since the width is proportional to the range) and that the cell rotates during the engagement to keep the width and depth oriented to the missile's velocity vector.

Continuation of the Engagement to Result

The above calculations are repeated every second until the ship and the missile are less than 1,000m apart. Then, the calculations are made every 100 ms. There will be a minimum ship-to-missile distance, after which the distance will start to rise. If the countermeasure fails, the minimum distance will be zero. A missile hit is reported out of the simulation. If the missile does not hit the ship, the minimum is reported as the miss distance.

Exercise

As an exercise for this engagement, calculate the location of all players at 3 seconds and record the results on the worksheet in Figure 7.15.

If these equations were set up on a spreadsheet, in Visual Basic, or in MathCAD, and repeated until a minimum distance from the missile to the ship were reached, they would determine whether or not the chaff was able to protect the ship.

Player locations
Ship location at 2 seconds _____ Ship velocity_____ Ship direction ____
Change in *x*_____ Change in *y*_____ Ship location at 3 seconds _____

Chaff location at 2 seconds _____ Chaff velocity ___ Chaff direction ____
Change in *x*_____ Change in *y*_____ Chaff location at 3 seconds _____

Missile location at 2 seconds _____ Missile velocity_____ Missile direction ____
Change in *x*_____ Change in *y*_____ Missile location at 3 seconds _____

Location of resolution cell centroid
Ship to chaff ΔX_____ ΔY_____ Distance_____

Ship RCS _____ Chaff RCS _____ Ratio of ship/chaff cloud RCS _____

Distance of resolution cell centroid from ship _____

Missile vector
Missile to centroid ΔX_____ ΔY_____

Angle from 90° to velocity vector _____ Missile velocity vector _____

Distance from missile to ship
Missile to ship ΔX_____ ΔY_____ Distance _____

Figure 7.15 Location of players at 3 seconds—worksheet.

8

Simulation for Training

The function of simulation in training is to provide a realistic operational experience by exposing trainees to the types of signals they will be expected to deal with under dangerous, real-world circumstances. Its principal value is to train the operator in a range of tasks that vary from simple "knobology" (i.e., What knob does what?) to the sophisticated use of EW equipment in extremely stressful situations. A secondary use of operator-interface simulation is to evaluate the adequacy of the operator interface provided by a system. (Are the system's controls and displays adequate to allow the operator to do the job that must be done?)

As with all training-related simulation, an EW simulator usually allows an instructor to control the difficulty of the student's training experience to meet the training objectives. Students typically progress from simple circumstances and tasks to complex ones, including the handling of system anomalies.

The only part of the situation (tactical environment, specific threats, friendly EW equipment) that matters in training simulation is what the operator actually sees, hears, and touches. No real signals or real equipment need be involved, although that is sometimes the best way to solve the training problem.

In the training role, a simulator is expected to provide signals no better and no worse than the student operator will see in the real world—within the limitations of the equipment controls and displays and the operator's perception limitations. Any time the output of a simulator is noticeably better or worse than the real world, negative training occurs.

In addition to supporting the student, a training simulator must support the instructor. The instructor must be able to determine what the student is doing, seeing, and hearing. This allows for corrective instruction and evaluation of student skills. The instructor needs to be able to call up scenarios of gradually increasing complexity, requiring the student to use increasingly complex operation-and-analysis capabilities.

A well-designed training simulator does not require either students or instructors to spend inordinate amounts of time learning to run the simulator. The instructors are supposed to be experts in EW, and any effort they spend learning to be experts in simulator defocuses them from their important core expertise. The students are on their way to becoming EW experts, so the same argument applies. Therefore, the simulator should "speak fluent EW instructor." That means that all simulator setup and operation commands should use the proper EW terms and units (rather than abstract digital commands) and allow the setup of scenarios in ways comfortable to professionals familiar with the way friendly and enemy assets are deployed and used.

8.1 Approaches to Training Simulation

8.1.1 Simulation Versus Emulation

There are two basic approaches to training simulation. One is to use the real equipment with simulated signal inputs. This is called *emulation*, because there are actual signals involved. The other is to simulate only the displays and controls that the operator sees and touches.

The use of the emulation approach is sometimes attractive because extra operational equipment is available, so no specific training simulation equipment is required. For EW systems for which there are only a few systems fielded, this may be the only practical approach.

Emulation has the advantage that all of the subtle effects caused by the operational equipment passing the signals will be present in the outputs to the operator. However, it has the disadvantage that real equipment must be maintained. Military equipment is expensive to maintain. It requires trained maintenance technicians, updated manuals, and adequate spares. Since the spares will probably be full operational quality, perhaps even mil-spec, they are also expensive. Another disadvantage of training on the real equipment is that it tends to be unavailable when it is needed for operations—wreaking havoc with training schedules. Finally, the real equipment usually does not permit the instructor to monitor student progress in a convenient way. We will discuss emulation further in Chapter 10.

Equipment simulating what the operator sees and touches is typically much less expensive to acquire and operate; it can also be designed as training equipment, with instructor-support features. However, the direct simulation of outputs requires careful attention to subtle equipment-related factors, to assure adequate simulation fidelity to support the training objectives.

8.1.2 Realistic Panels

Two approaches are commonly used in the provision of simulated operator controls and displays. One is to have control and display panels that look and feel like the real thing. The second is to use standard computer hardware to provide the control and display functions.

One obvious way to provide realistic panels is to use the actual control and display panels from the system, but to drive them directly from a computer, as shown in Figure 8.1. This approach has the advantage that the operator's training experience is realistic. The real knobs are where they should be, and they are the right size and shape. The real displays have the right level of flicker, and so on. This approach is becoming easier, given the trend to drive operational displays digitally. The real system controls and display are becoming more and more like computer terminals. Some actually are computer terminals. Thus, they are easy to drive by computer.

The real system displays are driven by injecting display signals in the form and format they would have if the hardware were being used operationally. Likewise, the signals coming from the operator control/display panels

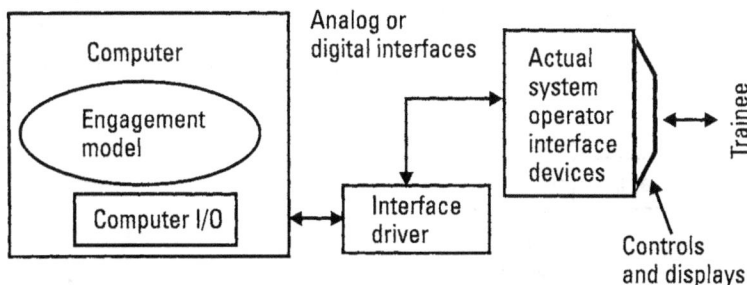

Figure 8.1 Operator-interface simulation with real system panel.

are sensed and converted to a form most convenient for the computer to accept.

There are two problems with using the real equipment panels: one is that such panels may be mil-spec hardware (with the attendant cost and maintenance issues); the second is that special hardware and software are often required to interface the panels to the computer. The signal conversions are not always implementable on standard computer I/O cards; sometimes an extra chassis (which needs to be built and maintained) is required. These considerations add cost to the procurement and operation of the simulator.

If real panels are not available, or if they do not meet the trade-off criteria, realistic panels can still be created. The simulated panel can have computer displays to simulate equivalent displays on the real panel, and can have switches or keyboard control inputs similar to the real controls and identically placed.

When a simulated control panel is used, the simulation computer must create display graphics formatted to match those of the operational equipment. The controls need to be sensed and their positions entered into the computer; Figure 8.2 shows the basic technique. Each switch provides logic-level "one" voltage to a specific location in a digital register when the switch is ON. The exact voltage depends on the type of logic used. Alternately, the switch may simply ground that location when it is ON. For an analog control (e.g., a knob) a shaft encoder is used. The shaft encoder typically provides pulses every few degrees as the control is moved, and the up/down counter converts those pulses into a digital-control-position word, which is input to the proper location in the register.

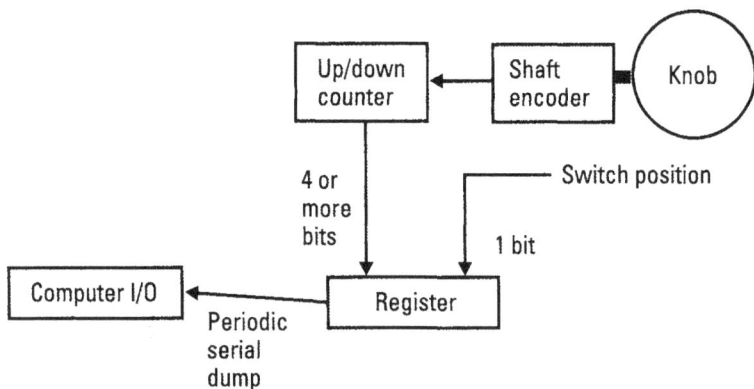

Figure 8.2 Computer interface to simulated panel controls.

The register is periodically read by the computer to sense the control positions. This is a very low-rate process (as computer speeds go) because of the low rate at which our hands move.

8.1.3   Depicted Controls and Displays

The other way to solve the problem is to use standard, commercial computer displays to simulate the operational displays. The controls can be created from commercial parts, or they can be simulated on the computer screen and accessed by keyboard or mouse, as in Figure 8.3.

Figure 8.4 provides an example. The figure shows the main computer screen for a training simulator for the AN/APR-39A radar warning receiver implemented on a desktop computer. The symbols on the screen move in response to simulated aircraft and mobile threat maneuvers. In this simulation, the control switches are also displayed on the screen. If the operator clicks on a switch with the mouse, the switch changes position on the screen and the system response to the switch action. Repeated mouse clicks on the audio control knob adjust the audio level.

8.2   Training Simulation Function

In most situations, it is practical to simulate a military engagement or equipment interaction of some kind, all in software. Then it can be determined what the operator would see, hear, and feel if he or she were in this postulated situation. It is also practical to sense what actions the operator takes and determine how the situation would be changed in response to those actions—and how that change would be sensed by the operator. The operator-interface simulator typically works from a digital model of the

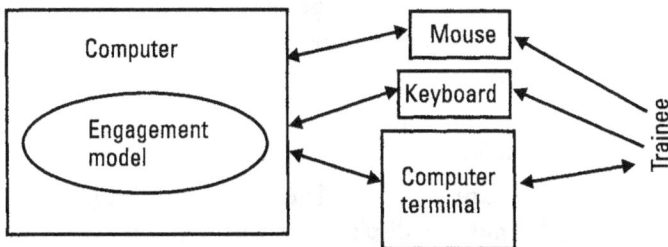

Figure 8.3 Depicted displays and controls.

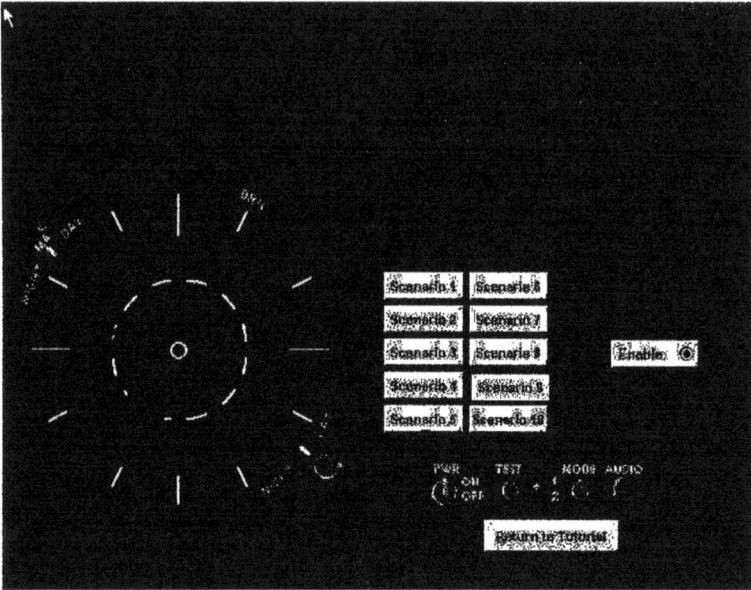

Figure 8.4 Example of depicted display and controls (Used with permission of I3C).

equipment and the engagement to determine the appropriate operator interfaces and present those to the operator.

If the operator actions are sensed in real time (with adequate fidelity) and the resulting situation is experienced by the operator in real time (with adequate fidelity), the operator will have the necessary training experience.

8.3   Required Fidelity

The fidelity necessary in a simulated operator interface is determined by a simple criterion: if the operator cannot perceive it, it need not be included in the simulation. The elements of fidelity are control response accuracy, display accuracy, and the timing accuracy of both.

8.3.1   Display-Time Fidelity

First let us deal with the time fidelity. The human eye requires about 42 ms to take in an image. Thus, if a display is updated 24 times per second (as in the movies), the operator perceives smooth action. In simulations bringing the operator's peripheral vision into play, the action must be somewhat quicker. Your peripheral vision is faster, so you will be bothered by flicker in your peripheral vision from the frames in a 24-frame-per-second presentation. In movies, which may be shown on wide-angle screens, the approach is to have 24 frames per second—but to flash the light twice for each frame so that your peripheral vision will not be able to follow the 48-per-second flicker rate.

Another perceptual consideration is that we perceive changes in patterns of light and dark (i.e., motion) much more rapidly than we perceive color changes. These two elements of a visual display are called *luminance* and *chrominance*. In video compression schemes, it is common practice to update the luminance at twice the chrominance update rate.

8.3.2 Control-Time Fidelity

A time-related consideration critical to simulation is the perception of the results of actions we have taken. Magicians say that the hand is quicker than the eye, but this is simply not true. Even the quickest hand motion (for example, pushing a stopwatch button a second time) takes 150 ms or more. (Try it to see how short a time you can catch on your digital watch!) However, you can perceive visual changes much faster. For example, when you turn on a light switch, you expect the light to come on immediately. As long as it actually comes on within 42 ms, your experience will be the same as the real-world situation. (See Figure 8.5.) In operator-interface simulation, the simulator must track binary switch actions and analog control actions such as the turning of a knob. While our perception of the results of a knob turning are less precise, it is still good practice to assume that the knob position should be translated into a visual response within the 42 ms for real-time fidelity.

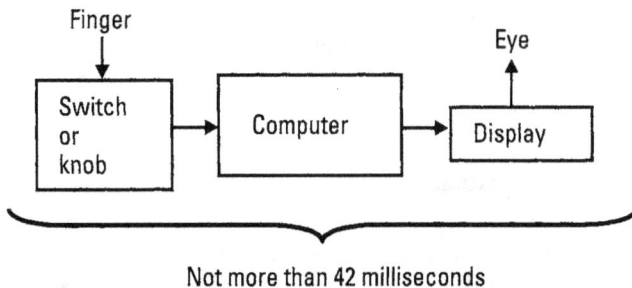

Figure 8.5 Hand-to-eye time fidelity.

8.3.3 Perceived Location Fidelity

Location accuracy is a little more tricky. We humans are not very good at perceiving absolute values of location or intensity—but we are very good at determining *relative* location or intensity. This means that if two items are supposed to be at the same angle or distance, we can sense a very small difference between their angles or distances. On the other hand, if both are off (together) by a few degrees or a few percentage points of range, we will probably not notice.

It is attractive from a training-effectiveness point of view to incorporate EW simulation into military vehicle simulators (primarily flight simulators, but also ships and land mobile systems). Operators can learn to deal with the realities of electronically equipped enemies while they operate the other subsystems associated with the military use of vehicles.

The only element that is added to the EW simulation when incorporated into a vehicle simulator is the coordinating of gaming areas.

A flight simulator, for example, has a visual gaming area and a radar land-mass simulator, among others. All of these gaming areas need to be coordinated so that the EW instruments indicate the same location for an EW threat as the visual displays will present to the operator—who may also be the pilot. There will be a specification for the required indexing of these gaming areas. For example, a threat location presented to the operator by two means might be required to be accurate to 20m.

Without this indexing, the operator would get conflicting cues, resulting in negative training.

Since the fidelity in a training simulator is not as critical as that required for a test-and-evaluation simulator, it may be practical to use lower-cost packaging techniques. One important consideration is spurious responses in signal generation. If spurs are perceptible in a training simulator, they may not impact the training value. In the real world, there are lots of extraneous signals passing through equipment. Therefore, the spurs can simply look like the real world if they are not high enough to dominate the system. Lower time fidelity can mean lower frequency digital logic, which can also reduce the requirement for expensive packaging techniques (for example, machined boxes for RF components).

There is a temptation to advertise simulators for use in both training and system testing. The danger is that the testing simulator might be too expensive for training use and the training simulator might be too low in fidelity to properly test the system. There have been some notable failures in attempts to satisfy both functions with one piece of simulation equipment, so approach this with care.

9

Simulation for Test and Evaluation

Simulation for T&E is used to test equipment under realistic signal environmental conditions. A standard piece of test equipment can be used to generate any type of signal. However, EW systems need to operate in the presence of a full signal environment, which usually contains many signals. But even with an unaffordable room full of single signal generators, you would still need to coordinate their operation to simulate the effects of a real-world military signal environment. T&E simulators provide that environment, in some appropriate form, to test receivers, subsystems, and whole systems.

In general, the frequency and location of the EW signal that must be received and processed in a limited amount of time is not known. Searching for that signal is made much more difficult by the presence of many similar signals that must be evaluated and rejected in some way. Hence, the T&E simulator tests the ability of both hardware and software to do their advertised job.

T&E simulators are by definition emulators. Since emulation is the subject of Chapter 10, it will not be specifically discussed here. The important issues related to T&E simulation are fidelity, entry points, and operator involvement.

Since T&E simulators are, in effect, specialized test equipment, they must support accurate testing. This means that they must not pollute the test data with test equipment inaccuracies. The general rule is that the simulator fidelity must be 10 dB better than the specifications to which the evaluated equipment is being tested. There are some circumstances in which more fidelity is required, and there are cases in which it is just too hard (or too expensive) to provide the 10-dB margin. Still, 10 dB is the standard, and is considered to deliver adequate test data in most circumstances.

In virtually every case, a T&E simulator has significantly higher fidelity than required for a training simulator. Ordinarily, a T&E simulator would support training fidelity requirements very handily—but its cost might be prohibitive for large-scale training applications.

The operator of a T&E simulator is typically a test technician who is skilled in the evaluation and troubleshooting of systems, subsystems, or components. A well-designed T&E simulator will allow him or her to set up and run tests using common technical terms and units. Special digital command words and the like will be minimized.

It is important that the setup of the simulator and the input and output conditions be easily captured. In most testing circumstances, it is necessary to be able to repeat the test exactly as it was run the first time.

Sometimes a T&E simulator needs to cycle through a lot of test steps. Being able to set up a test sequence ahead of time and let it run—perhaps unattended—is an extremely desirable feature.

Because of the fidelity requirement, it is normally necessary to use the same kind of construction techniques for a T&E simulator as that used in the equipment to which it supplies inputs. This means that RF components may need to be in machined housings, and so on.

Spurious responses in signal generation are far more serious in a T&E simulator than in a training simulator. While an operator can ignore spurs as extraneous signals, a receiver that is required to provide a high spurious-free dynamic range cannot be tested if the simulator has higher spurs (or even spurs close to the required isolation level).

In general, it is best to inject signals into a system under test immediately before the component or subsystem that is being evaluated. This means that the injected signal must represent real signals at that point in the system. The signals may have been converted in frequency, and will have certainly been either attenuated or amplified. The simulator injecting the signals is responsible for all of the system functions upstream of the injection point. If the signals are injected upstream of other components that precede the subsystem under test, it may be difficult or impossible to determine whether any anomalies seen were caused in the system under test or the upstream elements.

10

Emulation

Emulation involves the generation of real signals to be received by a receiving system, or by some part of that system. This is done either to test the system (or subsystem) or to train operators in the operation of the equipment.

In order to emulate a threat emission, it is necessary to understand all of the elements of the transmitted signal and to understand what happens to that signal at each stage of the transmission, reception and processing. Then, a signal is designed to look like the signal at some specific stage in the path. That signal is generated, and injected into the process at the required point. The requirement is that all of the equipment downstream of the injection point think it is seeing the real signal in an operational situation.

10.1 Emulation Generation

As shown in Figure 10.1, emulation, like any other type of simulation, starts with a model of what must be simulated. First, of course, the characteristics of the threat signals must be modeled. Then, the way the EW system will experience those threats must be modeled. This engagement model determines which threats the system will see and the range and angles of arrival at which the system will see each threat. Finally, there must be some kind of model of the EW system. This system model (or at least partial system model) must exist, because the injected signal will be modified to simulate the effects of all of the parts of the system that are upstream of the injection

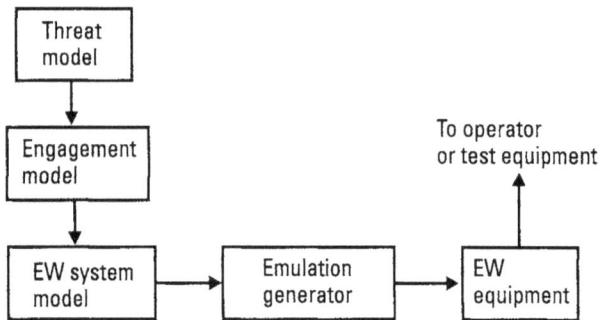

Figure 10.1 Elements of emulation.

point. The upstream components may also be affected by actions of down-stream components. Examples of this are automatic gain control and anticipated operator control actions.

10.2 Emulation Injection Points

Figure 10.2 is a simplified diagram of the transmission, reception, and processing path for a threat signal, and the points at which emulated signals can be injected. Table 10.1 summarizes the simulation tasks required by the selection of each injection point, and the following discussion expands upon the associated applications and implications.

Full Capability Threat Simulator (A)

This technique creates an individual threat simulator that can usually do everything that the actual threat can do. It is typically mounted on a carrier that can simulate the mobility of the actual threat. Since it uses real antennas like

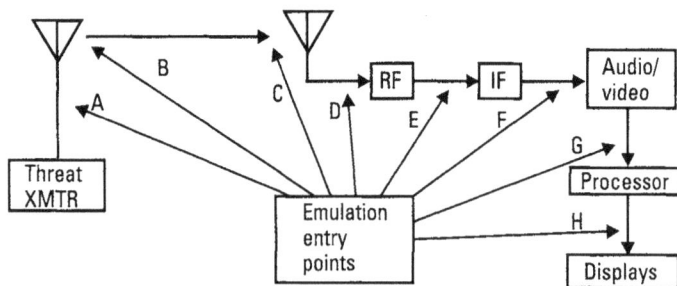

Figure 10.2 Emulation-signal injection points.

Table 10.1

Emulation Injection Points

Injection Point	Injection Technique	Simulated Parts of the Path
A	Full-capability threat simulator	The threat modulation and modes of operation
B	Broadcast simulator	Threat modulation and antenna scanning
C	Received signal energy simulator	Transmitted signal, transmission path losses, and angle-of-arrival effects
D	RF signal simulator	Transmitted signal, path losses, and receiving antenna effects including angle of arrival
E	IF signal simulator	Transmitted signal, path losses, receiving antenna effects, and RF equipment effects
F	Audio or video input simulator	Transmitted signal, path losses, receiving antenna and RF equipment effects, and effects from the selection of IF filters
G	Audio or video output simulator	Transmitted signal, path losses, receiving antenna and RF equipment effects, and effects from the selection of IF filters and demodulation techniques
H	Displayed signal simulator	The entire transmission/reception/processing path

those of the threat emitter, antenna scanning is very realistic; multiple receivers will receive a scanning beam at different times and at appropriate ranges. The whole receiving system is observed doing its job. However, this technique generates only one threat and is often quite expensive.

Broadcast Simulator (B)

This technique transmits a signal directly toward a receiver being tested. The transmitted signal will include a simulation of the scanning of the threat antenna. An advantage of this type of simulation is that multiple signals can be transmitted by a single simulator. If a directional antenna (with significant gain) is used, the simulation transmission can be at a relatively low power level and interference with other receivers will be reduced by the antenna's narrow beam-width.

Received Signal Energy Simulator (C)

This technique transmits a signal directly into a receiving antenna, normally using an isolating cap to limit the transmission to the selected antenna.

An advantage of this injection point is that the entire receiving system is tested. Coordinated transmissions from multiple caps can be used to test a multiple-antenna array such as a direction-finding array.

RF Signal Simulator (D)

This technique injects a signal that appears to have come from the output of the receiving antenna. It is at the transmitted frequency and at the appropriate signal strength for a signal from the antenna. The amplitude of the signal is modified to simulate the variation in antenna gain as a function of angle of arrival. For multiple antenna systems, coordinated RF signals are typically injected into each of the RF ports, simulating the cooperative action of the antennas in the direction-finding operation.

IF Signal Simulator (E)

This technique injects a signal into the system at an intermediate frequency (IF). It has the advantage that it does not require a synthesizer to generate the full range of transmission frequencies (since, of course, the system converts all RF inputs to the IF). However, the simulator must sense the tuning controls from the EW system so that it will input an IF signal only when the RF front end of the system (if it were present) would have been tuned to a threat signal frequency. Any type of modulation can be applied to the IF injected signal. The dynamic range of the signal at the IF input is often reduced from the dynamic range that the RF circuitry must handle.

Audio or Video Input Simulator (F)

This technique is appropriate only if there is some kind of unusual sophistication in the interface between the IF and the audio or video circuitry. Normally, you would choose injection points E or G rather than this point.

Audio or Video Output Simulator (G)

This very common technique injects audio or video signals into the processor. The injected signals have all of the effects of the upstream path elements, including the effects of any upstream control functions that are initiated by the processor or the operator. Particularly in systems with digitally driven displays, this technique provides excellent realism at the minimum cost. It also allows the checkout of system software at the minimum simulation complexity and cost. It can simulate the presence of many signals at the system antennas.

Displayed Signal Simulator (H)

This is different from the operator-interface simulation in that it injects signals into the actual hardware that displays to the operator. It is appropriate only when analog display hardware is used. It can test both the operation of the display hardware and the operator's (perhaps sophisticated) operation of that hardware.

10.3 Advantages and Disadvantages of Injection Points

In general, the farther forward in the process a signal is injected, the more realistic the simulation of EW system operation will be. A great deal of care must be taken if receiving-equipment anomalies must be accurately represented in a simulated signal. In general, the farther downstream in the process that injection is made, the less complex and expensive the simulation will be. In general, open-air-transmitted emulation techniques must be restricted to signals that are unclassified, so real enemy modulations and frequencies may not be used. However, techniques in which signals are hard-cabled to the EW system can use real signal characteristics for the most realistic possible testing of software and training of operators.

10.4 Emulation of the Receiving System

In Chapter 5 we discussed simulation of threats and what they look like when they arrive at the receiver location. Now we will investigate what those signals look like as they proceed through the EW system.

Everything upstream of the chosen emulation injection point must be emulated in order to properly create the signal to be injected. We will consider the emulation of the antenna function, the receiver function (in several stages), and the processing functions.

10.4.1 Receiving-Antenna Emulation

In an antenna emulator, the bore-sight gain (main beam peak gain) is recreated by increasing (or decreasing) the power from the signal generator, which is generating the RF signals as they arrive at the receiving site. Simulating direction of arrival is somewhat more complicated.

As shown in Figure 10.3, the direction of arrival of each received signal must be programmed into the simulator. In some emulator systems, a single RF generator can simulate several different non-time-coincident emitters. These signals are usually pulsed, but can be any short duty cycle signals. In this case, the antenna simulator must be told which emitter it is simulating (with enough lead time to set up its parameters for that signal). The antenna

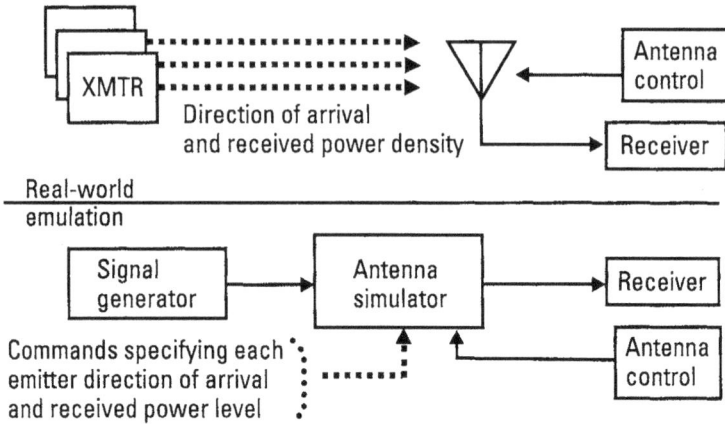

Figure 10.3 Antenna emulator function.

control function is not present in all systems, but when present, usually rotates a single antenna or selects antennas.

For directional antennas (as opposed to those which have fairly constant gain in all directions of interest) the signal from the signal generator is attenuated as a function of the angle from the bore-sight of the antenna to the simulated direction-of-arrival of the signal. The antenna orientation is determined by reading the output of the antenna control function.

You might like to refer back to Section 5.1, which discusses the two types of antennas used in the following examples.

Parabolic Antenna Example

Figure 10.4 shows the gain pattern of a parabolic antenna. The direction in which the antenna has maximum gain is called the *bore sight*. As the direction of arrival of an emitter moves away from this angle, the antenna gain (applied to that signal) decreases sharply. The gain pattern goes through a null at the edge of the main beam and then forms side lobes. The pattern shown is in a single dimension (e.g., azimuth). There will also be an orthogonal direction pattern (i.e., elevation in this case). The measured gain pattern can be stored in a digital file (gain versus angle) and used to determine the attenuation required to simulate any desired angle of arrival.

Although the side lobes in a real antenna will be uneven in amplitude, those in an antenna emulator are often constant. Their amplitude is lower than the bore-sight level by an amount equal to the specified side-lobe isolation in the simulated antenna.

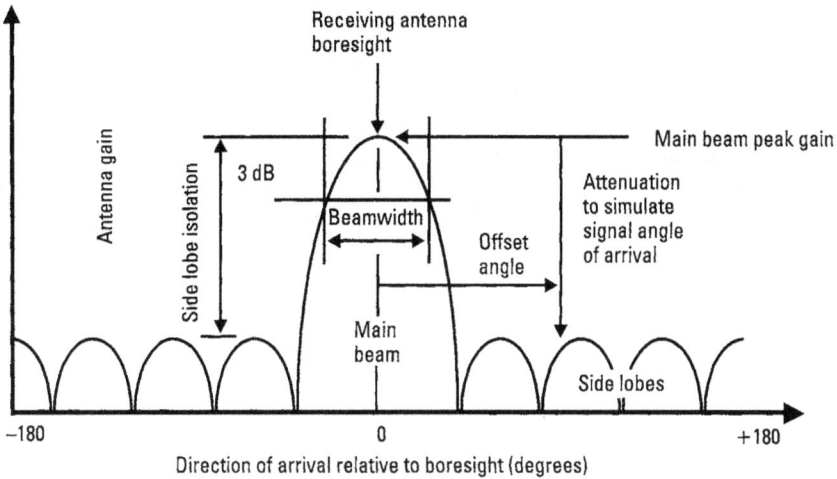

Figure 10.4 Typical antenna pattern for emulation.

If the emulation is to test a receiving system that operates with a rotating parabolic antenna, each of the simulated target signals would be input to the simulator (from the signal generator) at a signal strength that includes the antenna bore-sight gain. Then, as the antenna control rotates the antenna (either manually or automatically), additional attenuation is added by the antenna emulator. The amount of attenuation is appropriate to simulate the received antenna gain at the offset angle. The offset angle is calculated as shown in Figure 10.5.

RWR Antenna Example

The antennas most commonly used in radar warning receivers (RWRs) have peak gain at bore sight, which varies significantly with frequency. However, these antennas are designed for optimum gain pattern. At any operating frequency in their range, their gain slope approximates that shown in Figure 10.6. That is, the gain decreases by a constant amount (in dB) as a function of angle from bore sight to 90 degrees. Beyond 90 degrees, the gain is negligible (i.e., the signal from the antenna is ignored by the processing for angles beyond 90 degrees).

The complicated part of this is that the antenna gain pattern is conically symmetrical about the bore sight. This means that the antenna simulator must create attenuation proportional to the spherical angle between the

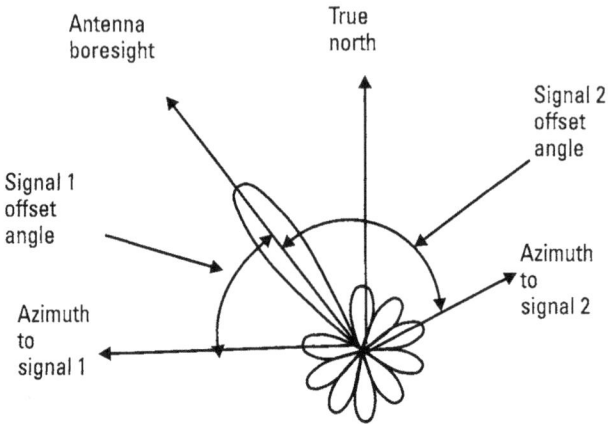

Figure 10.5 Antenna gain to multiple simulated signals.

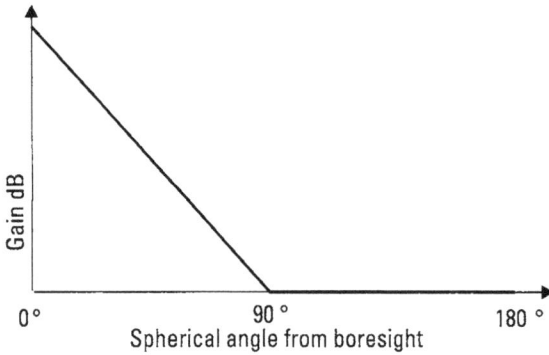

Figure 10.6 RWR antenna gain pattern.

bore sight of the antenna and the direction of arrival of each signal, as shown in Figure 10.7.

These antennas are typically mounted ±45 degrees and ±135 to the nose of the aircraft and depressed a few degrees below the yaw plane. Add to this that tactical aircraft often do not fly with their wings level, and thus may be attacked by threats from any (spherical) angle of arrival.

The typical approach to calculation of the offset angle is to first calculate the azimuth and elevation components of the threat angle of arrival at the aircraft's location, then to set up a spherical triangle to calculate the spherical angle between each individual antenna and the vector toward the emitter.

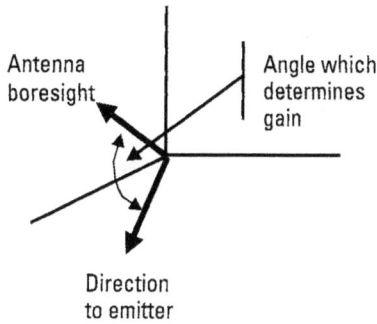

Figure 10.7 Antenna bore-sight-to-signal angle.

In a typical RWR simulator application, there will be one antenna output port for each antenna on the aircraft (four or more). The spherical angle from each threat to the bore sight of a single antenna will be calculated for each antenna output and the attenuation set accordingly (see Chapter 3).

Other Multiple Antenna Simulators

Direction-finding systems that measure the phase difference between signals arriving at two antennas (interferometers) require very complicated simulators—or very simple ones. To provide continuously variable phase relationships is very complex, because the phase measurements are very precise (sometimes part of an electrical degree). For this reason, many systems are tested using sets of cables that have the appropriate length relationships to create the correct phase relationship for a single direction of arrival.

10.4.2 Receiver Emulation

As shown in Figure 10.8, an RF generator can produce signals that represent transmitted signals arriving at the location of the receiver. An antenna emulator will adjust the signal strength of those signals to represent the action of the receiving antenna. Then the receiver emulator will determine the operator control actions and produce appropriate output signals as though a receiver had been controlled in that manner. In many modern systems, direct receiver control is performed by computers, but the actual receiver control functions are the same.

It is generally not useful to emulate only the receiver functions, but rather to include the receiver functions in a simulator (emulator) that represents everything that happens upstream of the receiver output. (Note that this includes the signal environment arriving at the receiving antenna and the

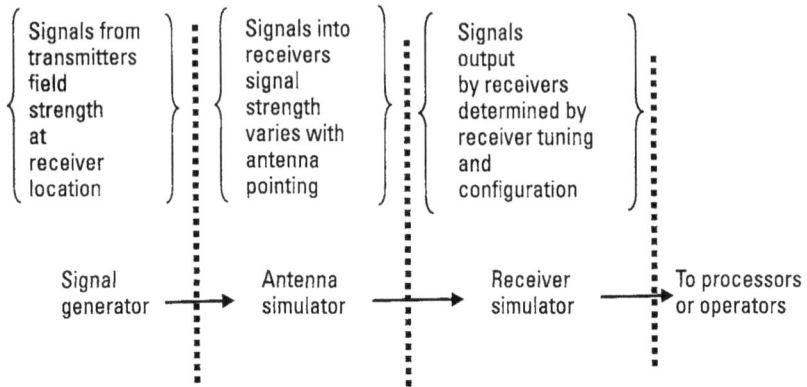

Figure 10.8 Stages of signal emulation.

function of the antenna.) Such a combined emulator is typically told (through digital inputs) the parameters of received signals and the operator control actions. In response to that information, it produces the appropriate receiver output signals.

The Receiver Function

Our purpose in this section is to consider the portion of that emulation which represents the receiver. First, let us consider the receiver function, apart from the mechanics of receiver design. At its most basic, a receiver is a device that recovers the modulation of a signal arriving at the output of an antenna. To recover that modulation, the receiver must be tuned to the frequency of the signal and must have a discriminator appropriate to the type of modulation on that signal.

In practical receivers there may be multiple outputs, and there are typically several control functions (either manual or automatic) to select the frequency, bandwidth, and modulation accepted by the receiver. A receiver emulator will accept values for the parameters of signals arriving at the receiver input and will read the controls set by an operator (or control commands from a computer). It will then generate output signals that represent the outputs that would be present if a particular signal (or signals) were present and the operator (or computer) had input those control operations.

Receiver Signal Flow

Figure 10.9 shows a basic functional block diagram for a typical receiver. It could be in any frequency range and operate against any type of signals.

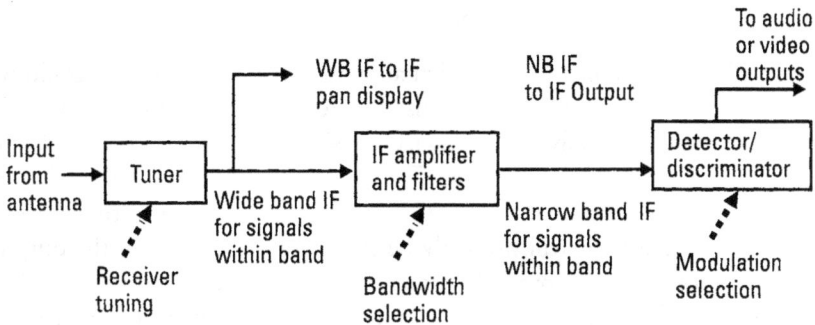

Figure 10.9 Functional flow of a typical receiver.

This receiver has a tuner that includes a tuned preselection filter with a relatively wide pass band. The output of the tuner is a wideband intermediate frequency (WBIF) signal, which is output to an IF panoramic (pan) display. The IF pan display shows all signals in the preselector pass band. The preselector is usually several MHz wide, and the WBIF is usually centered at one of several standard IF frequencies (455 kHz, 10.7 MHz, 21.4 MHz, 60 MHz, 140 MHz, or 160 MHz, depending on the frequency range of the receiver). Any signal that is received by the tuner will be present in the WBIF output. As the receiver tunes across a signal, the IF pan display will show that signal moving across the tuner pass band in the opposite direction. The center of the WBIF band represents the frequency to which the receiver is tuned, and signals in this output vary with the received signal strength. A skilled operator can recognize some kinds of modulation on signals in the IF pan display.

The wideband IF signal is passed to an IF amplifier that includes (in this receiver) several selectable band pass filters centered at the IF frequency. This postulated receiver has a narrowband intermediate frequency (NBIF) output—perhaps to drive a direction-finding or predetection recording function. The NBIF signal has the selected bandwidth. The signal strength of NBIF signals is a function of the received signal strength, but that relationship may not be linear, since the IF amplifier may have a logarithmic response or may include automatic gain control (AGC).

The NBIF signal is passed to one of several discriminators, selected by an operator (or a computer). The demodulated signal is audio or video. Its amplitude and frequency are not dependent on the received signal strength, but rather on the modulation parameters applied to the received signal by the transmitter.

The Emulator

Figure 10.10 shows one way in which an emulator for this receiver could be implemented. If the emulator were used to test a piece of processing hardware, it would probably be necessary to simulate the anomalies of the receiver—for example, band edge effects from mistuning and modulation outputs when the improper modulation is selected. However, if the purpose is to train operators, it will probably be sufficient simply to kill the output when the receiver is not properly tuned or the improper discrimination has been selected. This will make the simulator much simpler, and will most likely satisfy the training objectives.

Figure 10.11 shows the frequency and modulation part of the emulator logic for a receiver emulator designed for training. For convenience, we will call the frequency of a simulated signal SF and the receiver tuning frequency (i.e., the tuning commands input to the simulator) RTF.

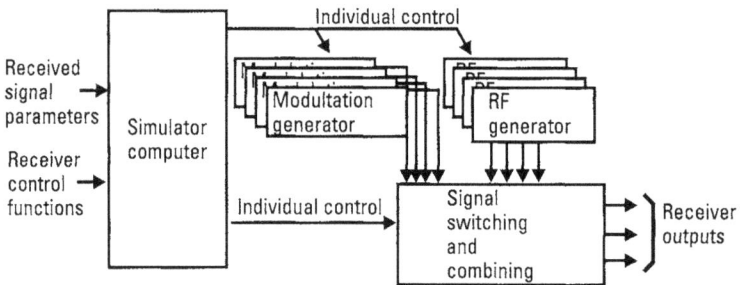

Figure 10.10 Block diagram of receiver emulator.

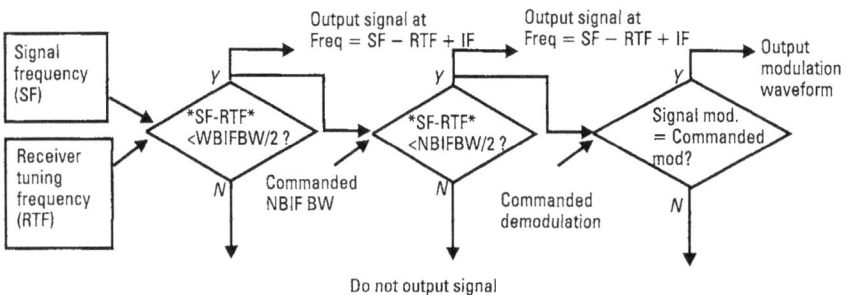

Figure 10.11 Receiver emulator logic.

For a signal to be displayed in the WBIF output, the absolute difference between SF and RTF must be less than half the wideband IF bandwidth (WBIF BW). Its output frequency will be

$$Frequency = SF - RF + IF$$

IF is the WBIF center frequency. This will cause the generated signal to move across an IF pan display in the opposite direction to the tuning. Note that this signal should have the proper modulation on it.

If the absolute difference between SF and RTF is less than half of the selected narrowband IF bandwidth, the signal will be present at the NBIF output. Its frequency will be determined by the same equation used for the WBIF output, but now IF is the center frequency of the narrowband IF.

Since this is a training simulator, the logic requires only that the modulation of the received signal match the demodulator that the operator has selected. If so, the operator will see (on a scope) or hear the modulation from the simulated signal. Otherwise, there will be no output.

Signal-Strength Emulation

The signal-strength input to a receiver depends on the effective radiated power of the signal and the antenna gain at its direction of arrival. As shown in Figure 10.12, the signal strength at each IF output is dependent on the net gain transfer function. In this receiver, WBIF output level is linearly related to the received signal strength because the gains and losses through the tuner are linear. The NBIF output is proportional to the logarithm of the received signal strength since the IF amplifier has a log transfer function.

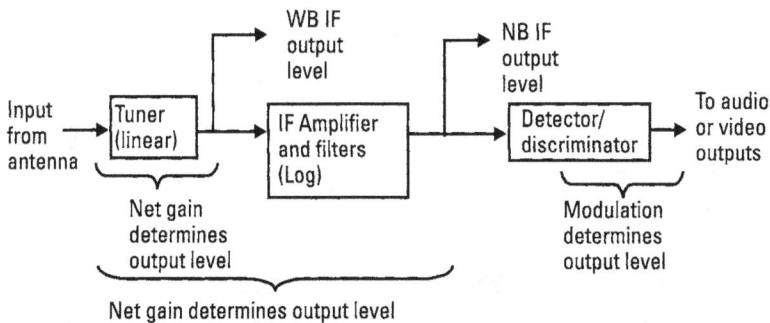

Figure 10.12 Levels of receiver outputs.

The amplitude of the audio or video output depends on the modulation level and the setting of the volume control.

10.4.3 Processor Emulation

In general, modern processors accept IF or video outputs and derive information about received signals (direction of arrival, modulation, etc.). This information is most often presented to an operator as computer-generated audio or visual indications. Thus, the emulation of the processing simply involves the generation of appropriate displays when the simulation calls for specific signals to be received.

10.5 Multiple-Signal Emulation

A characteristic of an EW threat environment is that many of the signals have short duty cycles. Therefore, it is possible to use a single generator to produce more than one threat signal. This has the advantage of significantly decreasing the cost per signal. However, as you will see, this cost saving can come at a performance cost. In this column, the last in the current simulation series, we will discuss the various ways to achieve multiple signal emulation.

The following discussion covers two basic methods for the emulation of multiple signals. The basic trade-off between the two methods is cost versus fidelity.

10.5.1 Parallel Generators

For maximum fidelity, a simulator is designed with complete parallel simulation channels, as shown in Figure 10.13. Each channel has a modulation generator, an RF generator, and an attenuator. The attenuator can simulate the threat scan and the range loss, and (if appropriate) the receiving antenna pattern. The modulation generator can provide any type of threat modulation, pulse, CW, or modulated CW. This configuration can provide more signals than the number of channels, since not all signals will be simultaneous. However, it can provide a number of instantaneously simultaneous signals equal to the number of channels. For example, with four channels, it could provide a CW signal and three overlapping pulses.

10.5.2 Time-Shared Generators

If only one signal needs to be present at any instant, a single set of simulation components (as shown in Figure 10.14) can provide many signals. This configuration is normally used only for a pulsed signal environment. A control

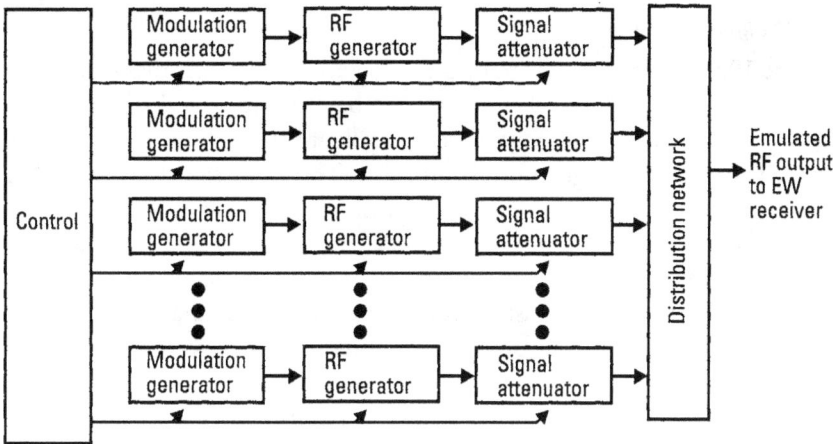

Figure 10.13 Parallel signal generators.

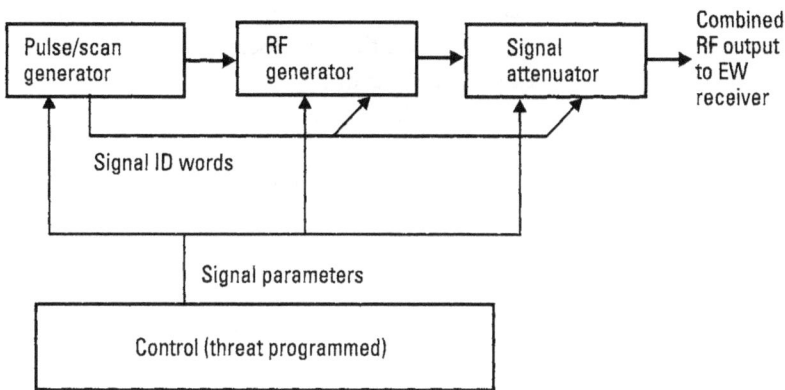

Figure 10.14 Shared component emulator.

subsystem contains the timing and parametric information for all of the signals to be emulated. It controls each of the simulation components on a pulse-to-pulse basis. The drawback of this approach is that it can output only one RF signal at any given instant. This means that it could output one CW or modulated CW signal, or any number of pulsed signals as long as their pulses do not overlap. As shown below, there is actually a restriction on pulses that closely approach each other in time, even if they do not actually overlap.

Pulse Signal Scenario

Figure 10.15 shows a simple pulse scenario with three signals that contain no overlapping pulses. All of these pulses can be supplied by a single simulator string controlled on a pulse-by-pulse basis. Figure 10.16 shows the combined video from the three signals on the first line. This is the signal that would be received by a crystal video receiver covering the frequencies of all three signals. The second line shows the frequency control that would be required to include all three signals in the RF output from the simulator. Note that the correct signal frequency must remain for the full pulse duration. Then, the synthesizer in the RF simulator has the interpulse time to tune to the frequency of the next pulse. Note that the synthesizer tuning and settling speed must be fast enough to change by the full frequency range in the shortest specified interpulse period. The third line of the figure shows the output power required to properly simulate all signals on a pulse-by-pulse basis. This means that the attenuator must settle at the correct level with the required accuracy during the minimum interpulse time. The change between pulses can be up to the full attenuation range. Depending on the simulator configuration, this attenuation can be only for the threat scan and range attenuation or can include receiving-antenna simulation.

Pulse Dropout

When a simulator cannot provide all of the pulses present in the signal environment, the missed pulses are called *pulse dropout*. This can cause inaccuracies in the evaluation of a system under test because it is stressed in an unrealistic way. The following discussion describes the reason for dropouts when a signal generator creates pulses for more than one signal.

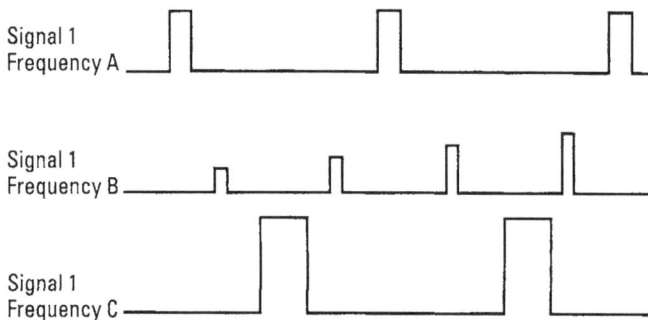

Figure 10.15 Simple pulse scenario with three signals.

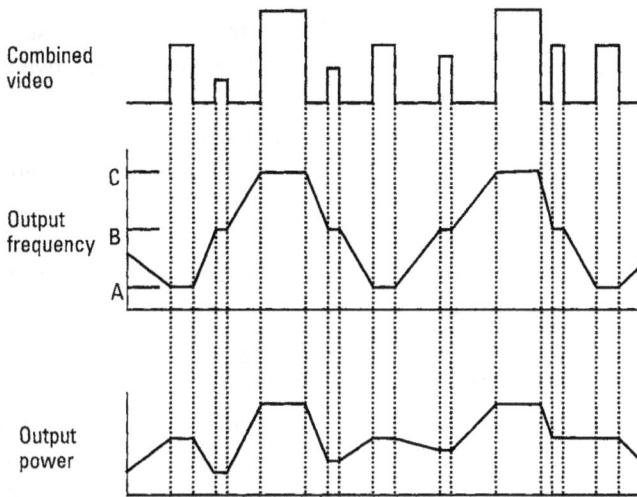

Figure 10.16 Combined signals.

Before the synthesizer and attenuator can start moving to the proper values for the next pulse, they must receive a control signal. This control signal, a digital word (Signal ID word), is sent before the leading edge of the pulse by an anticipation time, as shown in Figure 10.17. The anticipation

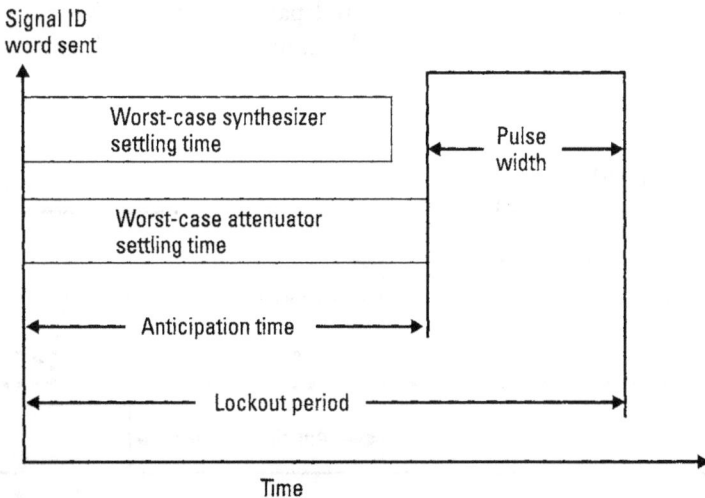

Figure 10.17 Required anticipation times.

time must be long enough for both the worst-case attenuator-settling time and the worst-cast synthesizer-settling time. The longer of the two times dictates the anticipation time. In the figure, the worst-case attenuator-settling time is shown as longer than the worst-case synthesizer-settling time. The lockout period is the time delay after the signal ID word before another signal ID word can be sent.

As shown in Figure 10.18, if a pulse occurs within a time that is the sum of the pulse width of the previous pulse and the anticipation time, it will be dropped from the simulator output.

Primary and Backup Simulators

The percentage of pulse dropouts can be significantly reduced if a second simulator channel is used to provide pulses that are dropped by the primary simulator channel, as shown in Figure 10.19. This assumes that the two simulator channels are well enough matched that the system under test will not detect a second signal. The analysis of the percentage of pulses dropped in various simulator configurations uses the Poisson equation (which is also useful in several other EW applications).

10.5.3   Approach Selection

The selection of the approach to providing multiple-signal emulation is a matter of cost versus fidelity. In systems requiring high fidelity and few signals, the choice is clearly to provide full parallel channels. Where slightly lower fidelity (perhaps 1% or 0.1% pulse dropout) is acceptable and there are

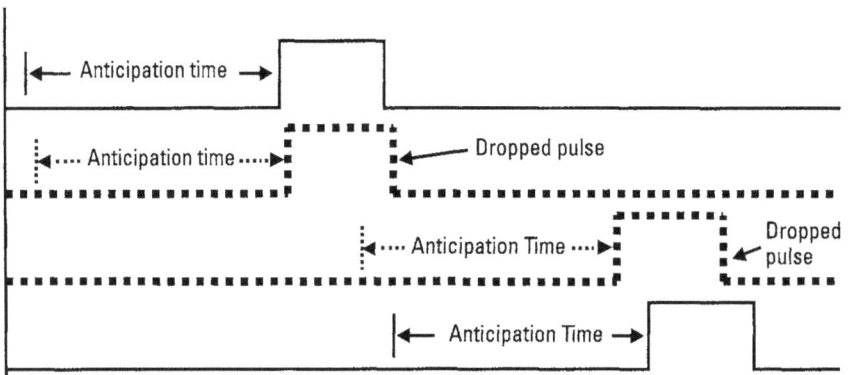

Figure 10.18 Dropped pulses because of lockout.

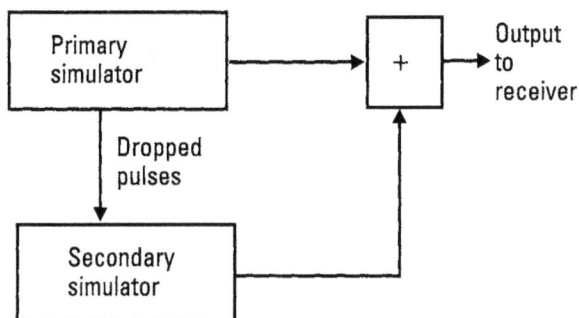

Figure 10.19 Primary and backup emulator channels.

many signals in the scenario, it may be best to provide a primary simulator with one or more secondary simulators. Note that the primary and secondary simulators must be matched closely enough that the system under test cannot distinguish them. Where pulse dropouts can be tolerated, a single channel simulator will do the job at significant cost saving. By setting priorities among signals, so as to avoid dropping pulses from higher-priority signals, the impact of dropped pulses may be minimized.

One approach that may give excellent results is to use dedicated simulators for specified threat emitters while using a single-channel multiple-signal generator to provide background signals. This tests the ability of a system to process a specified signal in the presence of a high-density pulse environment.

Glossary

This glossary is not intended to cover the electronic warfare field; it focuses on the special terms associated with the EW modeling and simulation field. The definitions are specific to this field.

Antenna scan   The angular motion of an antenna in its coverage of a spatial volume. In simulation, there is a strong focus on the time versus received power for threat emitter scans from the point of view of an EW receiver.

Emulation   The generation of signals in a form appropriate for the testing of receiving systems or parts of those systems. If the emulated signal is injected into the middle of an EW receiver, it represents what a transmitted threat signal would look like when it arrives at that point in the system—including the effects of the tactical situation and the status of all upstream hardware.

Emulation injection   An emulated signal is input to the threat generation, propagation, reception, processing stream at an "injection point." This point can be at the enemy transmitter, at the output of the enemy antenna, at the friendly receiver antenna, at the receiver, at the receiver's IF or video circuitry, at a digital processor, or at the systems displays.

Engagement   The series of events in which an enemy asset (e.g., a radar or missile) interacts with a friendly asset (a receiver, a platform, etc.).

Equipment anomalies Unintended effects of (usually friendly) equipment that cause the output to be other than intended by the designer. This can include spurious responses, data distortion, incorrect outputs from digital processors under certain operational conditions, and so on.

Fidelity The accuracy and granularity of the way a situation, event, or signal is handled in a model or simulation. Typical fidelity issues include the update interval in a model of an engagement, the number of pixels in a visual display, and the level of accuracy of an emulated signal.

Gaming area The area in which a series of events takes place during a modeled engagement. For a naval engagement, this can be a two-dimensional area of ocean. For an aircraft engagement, it will include the whole area in which the aircraft and missiles maneuver and the line-of-sight paths to any transmitters and receivers involved.

Modeling Modeling involves the mathematical characterization of an engagement, a process, or a piece of equipment. A model must be established before any simulation can be conducted.

Negative training Negative training occurs when a simulated system output is other than what would be present in the real situation being simulated. This becomes a problem when the inaccuracy (either better or worse than real life) is sufficient to impact the ability of a student to develop the desired skills from the training.

Operator-interface simulation The artificial creation of the control and display actions a trainee will experience in the performance of some task for which he or she is being trained. There are no real signals involved, only the controls and displays.

Operator interfaces Anything that the operator sees, hears, or touches during the operation of a system. This includes the way data is presented and the "feel" of controls.

Parallel generators When several signal generators are used—one to generate a simulated signal—they are said to be parallel. Parallel generators are used to minimize pulse dropouts or to generate multiple continuous signals.

Players Each of the participants in a modeled engagement is a player. Each is characterized by the way it moves, the signals it emits or receives, and the way that its movement or operational modes are dictated by the actions of the other players.

Point of view Every simulation is driven by a point of view. The events of engagement are portrayed in terms of what is perceived from that point of view. For example, a training simulator would typically have the point of view of the trainee. A simulator to test a system would have the point of view of the equipment under test. In both cases, complex interactions are converted to the generation of that which is perceived from the point of view.

Pulse Dropout When a simulator cannot produce every pulse of multiple signals presented to a receiver, the pulses that are not output constitute pulse dropout.

Signal Environment The totality of signals (friendly and enemy) that are encountered by a receiving system during an engagement.

Simulation Simulation has two meanings. One includes the whole field of modeling and simulation (i.e., operator-interface simulation, emulation, and computer simulation or modeling). The second use is as a substitute term for operator-interface simulation.

Simulator In EW simulation, this is a device that generates simulated signals in some way. It may generate actual signals in any form (broadcast, RF, video, etc.) or it may include artificial controls and displays that react as though the signals were present.

T&E Test and evaluation of a system, subsystem, or component using emulated signals.

Threat A threat is literally a weapon of some kind. However, it is also used to indicate a signal associated with a weapon system. This can include enemy radar and communication signals.

Threat environment The totality of the threat signals observed by a system during an engagement.

Threat scenario A series of threat signals of various types to simulate what a receiver or an operator would experience during an actual operational activity. The timing of the start and stop of each threat signal and the changes in the parameters of each are appropriate to the simulation point of view during the simulated operation.

About the Author

There should be a law against authors writing these author biography sections in the third person. Since few people read this section before at least thumbing through the book, you know by now that I have far too much respect for your intelligence to try to fool you that way. You know very well that some all-knowing other person did not write glowing prose about me in this hallowed place. It's me, Dave, your colleague.

What you need to know about me is that I have been involved in electronic warfare (EW) and EW simulation for a long, long time. After leaving the army, I spent the next 24 years designing receiving systems for EW and reconnaissance applications from submarines to space, at frequencies from just above dc to just above light, and managing the efforts of others (most smarter than I) who were doing the same. During this time, I ran the systems engineering group at Antekna, one of the earliest companies in the EW simulation business, for a couple of years. Over the years, I have also participated in simulation programs and testing programs using simulations in several companies in a variety of locations and have been the principal investigator on a wide range of EW and reconnaissance modeling projects.

I hold a BSEE from Arizona State University and an MSEE from the University of Santa Clara, both with majors in communications theory, and I have published about 100 articles on communications, EW, and EW-simulation-related topics in a variety of technical magazines and journals. I have also written extensive tutorial sections in various handbooks. I have never published an integral sign (except once, in a cartoon)—all of these articles and tutorial sections have explained communications, EW, and simulation phenomena using a practical application-oriented approach like that used in this book.

I now make my living as the owner of a small company that performs studies for the government and for prime defense contractors. I also teach short courses all over the world on communication theory, EW, and simulation-related subjects.

Index

processors, 126–27
receivers, 119–26
transmitters, 117–19
EW systems
injection points of emulated signals, 5
life cycle, 10
threat environment and, 4
Fidelity
control-time, 189
cost of, 8
dimensions considered for, 169
display-time, 188–89
engagement, 168–69
in EW simulation, 7–8
hand-to-eye time, 189
importance, 7
perceived location, 190
training simulation, 188–90
Fixed tuned receiver, 124
Frequency difference of arrival (FDOA),
136–37
defined, 136
isofreq contours, 137
with TDOA, 137
Frequency modulation (FM), 32–33

Gain, 105
adjustments vs. efficiency, 115
beam width vs., 114–16
nonsymmetrical, 116
pattern, 106
See also Antennas
Gaming area, 161–62
aircraft in hostile airspace, 171–72
construction steps, 161–62
defined, 161
illustrated, 163
ship attacked by antiship missile, 178
See also Engagement modeling
GPS guidance, 25, 26
Ground-based weapons, 140–42
acquisition, 141
fusing, 142
guidance, 142
launch, 141
search, 140
tracking, 141

Guidance, 22–26
active, 23, 24
command, 23
GPS, 25, 26
in ground-based weapons, 142
imaging, 25
passive, 24–25
semiactive, 23, 24
terrain-following, 25, 26
See also Radar

Hand-to-eye time fidelity, 189
Helical scan
defined, 150
illustrated, 151
observed power history, 151
See also Antenna scans
HF band, 30
High-power laser (HPL), 52
High-speed antiradiation missiles
(HARMs), 53
Horn antenna, 112

Imaging guidance, 25
Infrared countermeasures, 53–55
Instantaneous frequency-measurement
(IFM) receiver, 121
Interferometer DF, 133–35

Jamming, 43–52
barrage, 49
basic concept, 43
communications, 43, 44–45, 100–101
cover, 48–49
deceptive, 49–51
deceptive communications, 51–52
defined, 43
noise, 48
radar, 44, 45–46
self-protection, 47–48, 103–4
spot, 49
standoff, 46–47, 101–3
swept spot, 49
See also Electronic attack (EA)
Jamming-to-signal (J/S) ratio, 44–45,
99–104
communications jamming, 45, 100–101
defined, 99

Radio propagation, 83–104
  effective range, 93–95
  J/S ratio, 99–104
  losses, 86–90
  one-way link equation, 83–86
  radar detection range, 99
  radar range equation, 95–97
  range limitation from modulation,
    97–98
  receiver sensitivity, 90–93
Rain loss, 89–90
Range
  detectability, 20, 94
  detection, 20, 99
  effective, 93–95
  equation, 20, 95–97
  limitation from modulation, 97–98
  maximum unambiguous, 21
  minimum, 21
Range gate pull-off (RGPO), 50–51
Raster scan, 151–52
  defined, 151
  illustrated, 152
  observed power history, 152
  See also Antenna scans
Receiver emulation, 201–6
  emulator, 204–5
  receiver function, 202
  signal flow, 202–3
  signal-strength, 205–6
  stages, 202
  See also Emulation; Receiving system
    emulation
Receivers, 119–26
  channelized, 122–23
  compressive, 121–22
  crystal video (CVRs), 119–21
  digital, 123–24
  fixed tuned, 124
  instantaneous frequency-measurement
    (IFM), 121
  subsystems, 125–26
  superheterodyne, 124–25
  terrain-masking of signals to, 159
  types of, 119–25
  See also EW equipment

Receiver sensitivity, 90–93
  calculation example, 93
  components, 91
  CVRs, 120
  defined, 90
  kTB, 91
  noise figure, 91–92
  SNR, 92–93
Receiving system emulation, 197–206
  processor, 206
  receiver, 201–6
  receiving-antenna, 197–201
  See also Emulation
Right spherical triangle, 70–72
  defined, 70
  illustrated, 71
  Napier's rules, 70, 71
  See also Spherical triangle
Root mean square (RMS) accuracy, 131

Sampling rate, 81
Search radars, 21–22
Sector scan, 150, 151
Seduction chaff, 56, 57
Self-protection jamming, 47–48
  defined, 47, 103
  geometry, 47, 104
  J/S formula, 104
  signal strength formula, 103–4
  See also Jamming
Semiactive guidance, 23, 24
Ship attacked by antiship missile, 175–81
  defined, 175–76
  engagement, 178–79
  exercise, 180–81
  gaming area, 178
  player locations, 179–80
  players, 177–78
  point of view, 178
  See also Engagement modeling
Shipboard ES systems, 39–40
  antennas, 39–40
  block diagram, 39
  displays, 40
  operation, 39
  SEI processor, 39, 40
  See also Electronic support (ES)